W9-CGV-161

DISCARD

What Is Genetics?

The mysteries of heredity have always challenged people's curiosity. Today science has replaced legends and part-truths with firmly based knowledge. In this introduction to the subject of genetics the authors discuss in an easily followed progression the structure of the cell (the basic unit of life), chromosomes (the material involved in cell division), and DNA (the chemical substance that carries the hereditary code). The text is illustrated with diagrams and microscopic photographs. The authors analyze the relationship of genetics and environment and evaluate theories of race and intelligence as genetic factors. Drawing upon their professional knowledge they describe the latest practical application of this science in the area of genetic counseling. This book is scientifically accurate, highly readable, and filled with the excitement of fast-developing knowledge in a fascinating field.

Also by Jerry Bornstein

Unions in Transition

What Is Genetics?

BY **Jerry** AND
Sandy Bornstein

ILLUSTRATED

Julian Messner : New York

Published by Julian Messner, a Simon & Schuster
Division of Gulf & Western Corporation, Simon &
Schuster Building, 1230 Avenue of the Americas,
New York, N.Y. 10020.

Manufactured in the United States of America

Design by Irving Perkins

Fourth Printing, 1981

Library of Congress Cataloging in Publication Data

Bornstein, Jerry.
What is genetics?

Bibliography: p.
Includes index.
1. Genetics. I. Bornstein, Sandy, joint author.
II. Title.
QH437.B67 575.1 79-15170
ISBN 0-671-32952-9

This is for our daughters
Lisa and Danica
with love and joy

. . . and in memory of
their grandfather Bob
Kassel, who still lives
in them.

ACKNOWLEDGMENTS

We would like to acknowledge the cooperation of:

The W. B. Saunders Company for permission to reprint illustrations from Moore, Keith L. *The Developing Human, Clinically Oriented Embryology*, 2nd Edition, © 1977 by the W. B. Saunders Company, Philadelphia, Pa.

Richard Lemon, Editor of the *Sunday News Magazine* for permission to reprint portions of the article, "Mariana's Healthy Baby," by Jerry Bornstein (Copyright 1976 New York News Inc. Reprinted with permission), which served as the basis for Chapter 15.

Harcourt Brace Jovanovich, Inc. for permission to quote from *The Biology of Race* by James C. King, copyright © 1971 by Harcourt Brace Jovanovich, Inc.

Doubleday & Company for permission to quote from *Roots* by Alex Haley, copyright 1976 by Doubleday & Company.

J. B. Lippincott Company for permission to reprint an illustration from *Your Heredity and Environment*, by Amram Scheinfeld, copyright © 1965, 1950, 1939 by Amram Scheinfeld.

We would also like to acknowledge our debt to the writings and lectures of Dr. James C. King of the New York University Medical School, which provided us with valuable insights and illustrative examples.

A special word of appreciation must go to Dr. Madelyn Milchman, a good friend and psychologist, who graciously donated time to help us understand some psychological concepts with which we were unfamiliar when we began work on the chapter on genetics and intelligence. She is not responsible for the views expressed in that chapter, but it would have been difficult for us to have come to grips with the material without her help.

—Jerry and Sandy Bornstein

Contents

Introduction

Human beings are the most sophisticated form of life nature has developed, and humans share this planet Earth with some 1.5 million other life forms, from the single-cell micro-organisms to insects, plants and animals. All these living things, from the simplest to the most complex, from the smallest to the largest, have one thing in common—they possess the unique ability to reproduce themselves and keep life going, even after the individual organism is dead. Individual living things live and die, but the species, or group to which they belong, generally continue generation after generation.

True, sometimes species become extinct—perhaps because of natural upheavals in the environment, like the coming of an ice age, or because of the thoughtlessness of people, both of which may disrupt or destroy the conditions of life. Sometimes species undergo change, evolving into still newer species at an excruciatingly slow pace, perhaps over millions of years. But in general, living things give rise to other living things like themselves.

How do species reproduce themselves generation after generation? What is inherited from the parents and what is not? What is passed on biologically from the parents and what results from environmental influences? How does the process of heredity work? What are the sources of variety? Genetics is the science that seeks to answer these questions and unravel the mystery of life.

Reproduction and heredity are closely intertwined. It is impossible to talk about genetics without discussing reproduction, but they are not the same thing. Reproduction is the physical process by which a new individual is formed, developed and born. Heredity is the way in which the information that determines the uniqueness of the individual being created is transmitted.

Heredity is as old as life. A billion and a half years ago the first living cell duplicated itself, split in half, producing the very first "daughter cells" and thereby created a new generation—the first new generation in history. Living things can be far more complex these days, but the *cell* is still the basic unit of life, and it is in the cell that the rudiments of heredity may be observed. Organisms may be composed of a single cell or billions of cells, but the processes by which they reproduce are not as dissimilar as they might superficially appear. In that very first cell existed the genetic material which had the potential to change and evolve into dramatically different life forms, up to and including human beings.

People have probably always wondered where they came from and how it happened, but it has taken quite a long time to understand the way cells function and life is passed on. Heredity may be as old as life, but genetics is a young science, dating back only to the beginning of the twentieth century. Now, every year brings dramatic breakthroughs which sweep aside long-standing theories and commonplace notions that may have been widely accepted just a few short years ago. Books written on genetics ten, even five, years ago may be outdated today.

Because it is so new as a science, genetics is not very far removed from the influence of popular mythology. Folk tale explanations were cast aside for many other sciences long ago, but still often persist for genetics. Many myths about heredity that were widespread in the days of your parents and grandparents have not been completely discarded. When these misconceptions are not pure fiction, they are often gross oversimplifications which obscure the beauty and intricacy of the hereditary process. In the pages that follow you may learn that some of the things you have read in the popular press, or heard from teachers or from your parents, are less than accurate. In a few years, it may very well be that some of the things we write today will no longer represent the latest scientific knowledge.

This book is intended as an introduction to genetics, the study of the hereditary process. It will acquaint the reader with what heredity means and how it works. It will present heredity in an historical context and explore the ethical and moral questions that the rapid developments in genetics have posed for human society. The glossary of unfamiliar terms at the back of the book will help the reader understand the scientists' language.

Genetics is a science that applies to all forms of life. Fruit flies and rats are frequently used laboratory animals in genetic research, but we think people are much more familiar to the average reader. As much as possible, we will discuss heredity in terms of human beings.

—Jerry and Sandy Bornstein

CHAPTER ONE

What Our Ancestors Thought

Curiosity has been part of human nature since the very beginning. People have always been able to see what was happening in the world around them and felt the need to explain it. When primitive society invented gods who were said to control the weather or the success of the hunt or the harvest, it was an attempt to provide an answer to the question: Why do things happen the way they do?

Unfortunately, curiosity alone is never enough to supply accurate answers to complicated questions. No matter how inquisitive or intellectually gifted people are, their ability to understand the world is limited by the development of human culture and technology at that precise historical moment. The intellectual contributions of any person must be based on what's been achieved before and on the tools available to search for new answers. If Einstein had been a caveman, he wouldn't have formulated the theory of relativity, but he might have had a hand in the invention of the wheel.

Although genetics is a new science, people have long been curious about heredity and reproduction. From far back in recorded history people have observed that reproduction follows a pattern. They noticed that children resemble their parents, that species of animals produce the same species and that plants give rise to plants similar to themselves. Through trial and error they learned they could manipulate the reproduction of plants for agricultural purposes.

It wasn't too long before people began to spin theories to explain these observations and account for the apparent transmission of heredity information between parent and child. In general the reproduction theories which preceded the modern era fell into two broad categories: *particulate* theories, which were by and large the product of ancient civilization, and *preformationist* theories, which were the product of the scientific revival of the Renaissance.

The first known theories of reproduction developed in ancient Greece. These particulate theories were based solely on observations that could be made with the naked eye. The technology of the period did not yet include the microscope, so that microorganisms and cells were unknown. Since reproduction occurs on a cellular level, the ancient theories were inevitably doomed to be inaccurate and distorted. Particulate theories suggested the transfer of information from each and every part of the parents' bodies to the offspring in order to create a new organism. Information from the leg, or the liver, or the heart, or the brain had to be transported to the offspring in order for it to develop the corresponding parts in its own body.

The preformationist theories of the Renaissance likewise reflected the gains and limitations of scientific knowledge and technology of the period. Cells had by now been discovered, but were only partially understood. The newly invented microscope greatly expanded the range of human vision, but the microscope's primitive nature limited the accuracy of observation. In one form or another, preformationist theories held that the parent had in his or her body specialized reproductive cells that contained within them whole or preformed offspring. All that was necessary was that these seeds be planted in the appropriate environment —the womb—to grow into a new organism.

Both types of theories were based upon and limited by the observation possible at the time. Some of these theories that we are going to examine may seem strange or even silly today, but they represented the best possible explanations based on the information available. In their own way, they attempted to explain the nonrandomness of reproduction and the transmission of information between parent and child during the creation of new life. Today we know that babies come from the fusion of *gametes,* or reproductive cells, in which the male's sperm cell fertilizes the female's egg cell. Modern technology has permitted the detailed study of this fertilization process, but it wasn't too long ago that even the most brilliant thinkers in the world hadn't the slightest inkling that human eggs or sperm existed.

BLOOD AND VAPORS

The ancients lacked the technical means to find accurate answers to the mysteries of reproduction, but they tried—and with some success. Through the painstaking efforts of trial and error, and careful observation, early societies learned about plant reproduction. Early Chinese records indicate that farmers were able to develop superior varieties of rice over six thousand years ago. An ancient Egyptian painting depicts men cross-pollinating date palms, indicating they must have understood something about sexual reproduction in plants. An ancient Babylonian tablet displays the pedigree of a family of horses through five generations, with carefully compiled information about height, length of mane, and other traits. This demonstrates an awareness that there was a correlation between parentage and characteristics.

It was the Greeks, however, whose civilization flourished five centuries before Christ, who made the first serious attempts at drawing up a coherent explanation of reproduction in animals and people. The details varied, but the thrust of these theories was usually quite similar. Each part of the parents' bodies secreted a substance which mixed and recombined during intercourse to form the corresponding parts of the offspring's body. Sometimes this substance was said to be vapors, semen or blood, and recombination was described as a simple mixing process or some form of coagulation of fluids. But the essentials were the same.

The men who suggested these theories were philosophers, not scientists. Intellectual pursuits were not yet divided into separate disciplines, and most Greek scholars dabbled in whatever interested them. Their theories were based on observations, but usually very superficial ones.

Aristotle was different. Like his peers, he wrote on many subjects. But when it came to biology, he alone understood the absolute necessity for systematic observation. He was the first real scientist in history. His efforts earned him the title of Father of Biology. To Aristotle, seeing was believing. He devoted many hours to a close examination of animal life. He compiled lengthy, detailed descriptions of what he saw, and classified and arranged his findings in logical order. So extraordinary were his powers of observation that he was able to see that dolphins were not fish but mammals. He noted that though they looked like fish and lived in the sea, dolphins were warm-blooded and had lungs; if they did not come to the surface to breathe, they would drown. It would take the rest of the world two thousand years to catch up to Aristotle—dolphins, porpoises and whales were being classified as fish until after the Renaissance in Europe.

In astronomy, Aristotle's keen observations led him to the conclusion that the world was round 1800 years before Columbus. Most people of Columbus' time thought he risked falling off the edge of the world by sailing west across the Atlantic.

But when it came to reproduction, there was too much the Father of Biology could not see with his eyes to enable him to be even remotely accurate. Aristotle advocated a theory in which blood was the key ingredient. In fact, for Aristotle, blood was the key to the entire human body. He saw the brain, for instance, as a special organ designed to cool the blood. He believed that a purified blood of the male was introduced into the female's body during intercourse and organized the less purified blood of the female into an embryo in the womb.

The Aristotelian misconceptions about blood and heredity continued for centuries. Expressions like "pure blood," "blood relative," or "blood line" which persist today are testimony to Aristotle's long-lasting influence. But, however much parents may speak of their children as their own flesh and blood, neither is a transmission belt of heredity.

Greek science petered out around 200 B.C. amidst war and the general decline of Greek civilization. Greece fell under the domination of the Roman Empire and philosophic inquiry shifted from natural science and mathematics to ethics and moral philosophy. The startling scientific achievements came to an end. When Rome in its turn fell, Western civilization entered a barren period called the Dark Ages. Science surrendered to superstition, and the insights gained by ancient Greeks faded into obscurity for a thousand years.

Christianity became the dominant religion. The revealed truth of the Bible, not the empirical observations of mortals, was seen as the source of knowledge.

During these long centuries of intellectual darkness, it fell to Moslem Arabs to keep the contributions of classical Greece from being lost forever. Ironically, Christendom derided the Moslems as infidels and barbarians. Nevertheless, Moslem scholars followed Arab armies as they swept across northern Africa, and into Sicily and Spain. They marveled at the treasures of Greek civilization that they found. Translating the works of Aristotle and others, they didn't add much in the way of new knowledge but they kept classical learning alive until the West was prepared to rediscover its heritage.

PREFORMATION AND ENCAPSULATION

The stifling effects of the Dark Ages were shaken off in the Renaissance. A cultural rebirth flowered in literature, the arts and in an energetic approach to the natural sciences—including the study of reproduction.

One late Renaissance scientist decided to test the old Greek theory of blood coagulation. William Harvey (1578–1657) mated twelve female deer and waited a few weeks before killing and dissecting one of the pregnant animals. He searched fruitlessly for evidence of coagulating fluids. Instead, he found a tiny embryo which didn't even resemble a deer at all. As the pregnancies progressed, the embryos he discovered in each dissection increasingly resembled a fawn. The embryo obviously was not formed by coagulation but developed gradually during the gestation period.

Harvey's observations led him to advocate a theory he called *ex ovo omnia,* which is Latin for "all from the egg." According to this theory, the embryo developed step by step in a steady progression from the egg of the mother. Harvey had never seen a human egg, but he believed that such a thing must exist. His theory marked a step forward, but he failed to see that fertilization was a necessary part of the process. Sperm had not yet been discovered, but Harvey overlooked the fact that he had made sure to mate his experimental deers in the first place. His ideas incorrectly denied the males of the species any role in reproduction. Harvey's belief that the egg itself merely grew into the new organism contained within it the basic elements of the preformationist theories that would be more pronounced in the next century, especially after the discovery of the sperm cell.

By the end of the seventeenth century, scientists had an amazing new tool at their disposal—the microscope. Aristotle would have been enthusiastic. The microscope enabled people to study things previously invisible to the naked eye. There was more to see and hence more to believe. Renaissance lenses were quite primitive compared to modern microscopes, but were sufficient to open up whole new areas to systematic scrutiny.

The next giant step was taken by a lensmaker named Anton van Leeuwenhoek who developed a flawless glass lens in the 1680s. A young medical student brought Leeuwenhoek a sample of a fluid discharged from a patient suffering from venereal disease. The lensmaker and the medical student discovered tiny, one-cell organisms, shaped like tadpoles with a solid, pointed head and a long, thin tail swimming around in the fluid. (Figure 1) The medical student thought the little creatures were parasites that had something to do with the patient's illness. But Leeuwenhoek suspected they were far more significant. His curiosity led him to study these little organisms more closely. Before he was done he had become the first person in history to observe living sperm in the semen of animals, fish, frogs and man. Since fish and frogs have external fertilization of the eggs, outside the

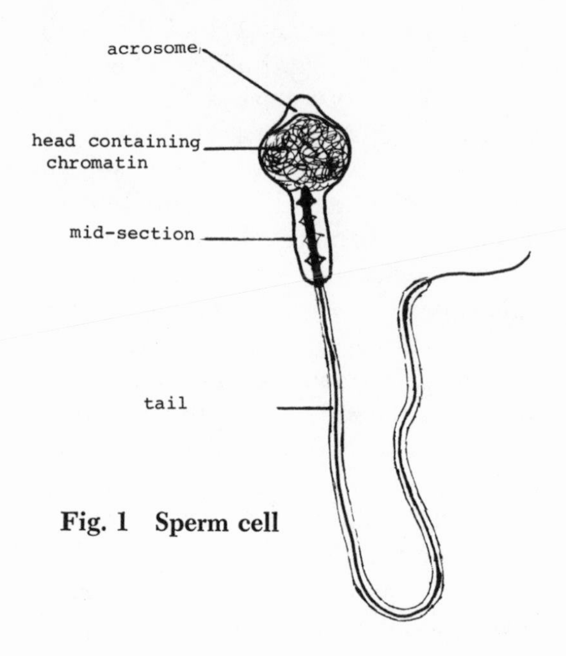

Fig. 1 Sperm cell

female's body, Leeuwenhoek noted that the males deposited their sperm over the eggs after the females laid them, and concluded that the sperms must have something to do with the formation of the embryo. (Figure 2)

The early microscopes were not powerful enough to make everything absolutely clear. This proved to be a serious problem when it came to something as tiny as the sperm cell. (Sperms are the tiniest cells produced in the human body. The female egg cell is only 1⁄100 of an inch in diameter, but it is still 85,000 times larger than a single sperm cell!) When scientists of the early 1700s studied sperm they noticed black squiggly lines within the pointed head of the cell. As if to prove the old adage that a little bit of knowledge can be dangerous, they began seeing things that just weren't there. Writer Isaac Asimov compares what happened next to the day long ago when primitive man looked up and saw the

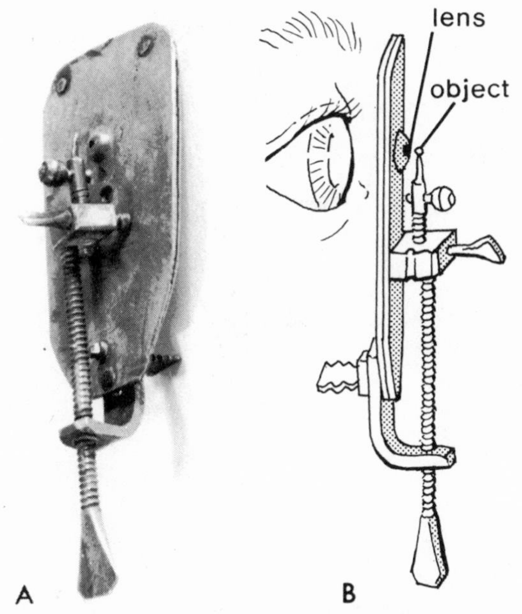

Fig. 2 A, Photograph of a *1673 Leeuwenhoek microscope.* B, Drawing of a lateral view illustrating its use. The object was held in front of the lens on the point of the short rod, and the screw arrangement was used to adjust the object under the lens. (Reprinted from Moore, *The Developing Human.*)

moon's pockmarked surface and imagined it was the face of the man in the moon. It began with a Dutch scientist who announced the discovery of a tiny little man in the head of a sperm and produced a sketch of the little fellow as proof.

This little man in the sperm was named *homunculus* (Latin for "little man"), and before long biologists everywhere were seeing miniature people under their microscopes. (Figure 3) The whole homunculus business began to get entirely out of hand, as a controversy erupted between those who

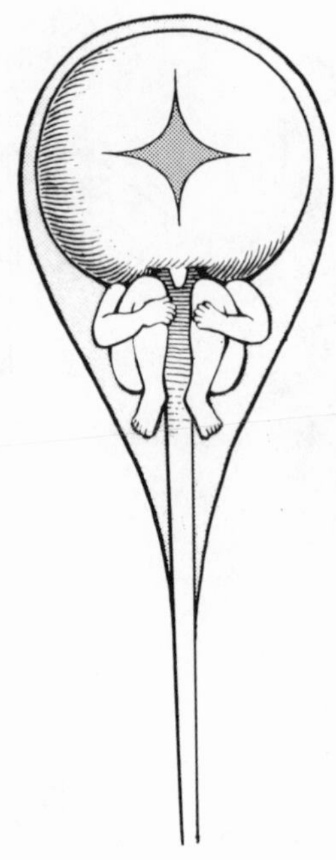

Fig. 3 Copy of a seventeenth century drawing by Hartsoeker of a sperm. The miniature human being within it was thought to enlarge after it entered an ovum (Reprinted from Moore, *The Developing Human.*)

insisted that the homunculus held his arms straight down at his sides and those who were adamant that he held them folded across his chest. Miniature roosters, frogs, and fish were soon being discovered in the sperm of their respective species.

The discovery of the homunculus prompted a new reproduction theory, the preformationist theory mentioned earlier. It was said that the male's sperm supplies the embryonic seed, which is actually a preformed, miniature human being. The female's uterus merely provides a warm, nourishing environment for the seed to grow in until birth. The preformationist theory was the flip side of Harvey's "all from the egg" theory. This time it was all from the sperm.

The discovery of miniature people in the sperm and the theories derived from their supposed existence were not universally accepted. There were those who denied that sperms had anything whatsoever to do with reproduction: the view that sperms were parasites infecting the male's body persisted until the 1840s. Some of the skepticism was better founded. Leeuwenhoek, himself, complained that he never saw an homunculus. One witty critic of preformation issued a report that he had seen an homunculus taking off his sweater. The sharpest criticism came from the members of a school of thought who championed the egg against the implicit glorification of the sperm in the preformationist theory. Harking back to "all from the egg" theory, one French scientist argued that it was the female egg that contained the seed of life—the male of the species had nothing to do with it. Citing a medieval quotation about "life dropping from the sky," he suggested that airborne spores entered the female body and triggered the life-bearing process.

This vigorous reassertion of the egg cell soon evolved into a full-blown theory of encapsulation, which held that the egg contained not miniaturized people, but tiny particles that had the power to form embryos. Like a series of Chinese boxes fitting one inside the other, each of these particles contained within it other particles that would someday create future generations. Encapsulationists now debated among themselves as to the exact number of eggs encapsulated in Eve's first egg cell in the Garden of Eden. Eventually the supply of life-giving particles would run out and human beings would face extinction.

Whatever its superficial differences with the preformationist theory and its conception of the homunculus, encapsulation shared the basic premise of its rival—that the body carried specialized cells that already possessed all the information necessary for the production of new life.

PREMODERN THEORIES

The theories of preformation and encapsulation battled with each other for decades at the same time that still other scientific investigators attacked both theories and began making inroads toward what we now recognize as the accurate explanation of reproduction. For example, experimental work in animal breeding, as well as detailed studies of family pedigrees, convinced Frenchman Pierre Louis de Maupertuis (1688–1759) that it was absolutely impossible for either the egg or the sperm to have a monopoly on inheritance. Offspring, he noted, always displayed at least some characteristics of each parent. A child might have a nose resembling his father's and the hair color of his mother. De Maupertuis believed that tiny particles migrated from various parts of the body to the reproductive organs where they were gathered and passed on to the offspring during intercourse. Reflecting the obvious influence of the Greek particulate theories, he believed that these particles remembered and reproduced the organs of their origin in the embryo. Because of his own empirical observations, de Maupertuis concluded that particles from one parent might dominate over the matching particle from the other parent, explaining why the child might have the mother's hair color and not the father's. In reaffirming that both parents play a fundamentally important role in reproduction and in suggesting for the first time that the hereditary contributions of one parent might exert canceling effect on the other parent's, he had taken a step in the right direction.

From his careful study of the development of a chick

embryo from an undefined blob to a baby chick pecking its way out of the shell, Kaspar Wolf (1731–1794) concluded that preformation was impossible. Organisms developed in a completely logical manner from generalized living tissue which gradually became specialized tissue and body organs. Wolf speculated that both male and female gametes—the reproductive cells—contained within them substances that became organized into body organs after fertilization. Other researchers before him had proved that mammals too had eggs, and therefore Wolf believed his ideas had universal applicability. He was close. Very close.

From the beginning, explanations of reproduction were based on what people could observe. The Greeks had no idea of the existence of cells, reproductive or otherwise. They saw that children resembled their parents and were forced to imagine that parts of the body secreted something that communicated to the offspring precisely how it should organize its body structure. Scientists of the seventeenth and eighteenth centuries knew about the existence of the cell, including sperm and egg, but this knowledge was fragmentary and disjointed. In their headlong rush to utilize this new knowledge, they scrapped some of the valid insights of the ancients, particularly the conception of the biparental nature of reproduction.

It wasn't until 1838 that several hundred years' worth of seemingly random empirical observations about the nature of cells was synthesized into a coherent theory. Thus, it was only 140 years ago that the cell was recognized as the basic unit of life, that simple one-cell organisms and complex creatures, like humans, were acknowledged as sharing the same basic building block of life. The realization that one-cell and complex organisms had much in common meant that the rudiments of heredity and reproduction could be analyzed on the level of the single cell.

Humanity now stood on the threshold of tremendous

progress in the life sciences. Scientists would now grapple with the cell, examine its constituent parts, and explore the world of microorganisms. Darwin would burst on the scene with his theory of evolution aimed at proving the basic unity of all life. The mystery of heredity, however, would wait until the beginning of the twentieth century to be solved.

CHAPTER TWO

Mendel and His Pea Plants

In 1865, a plump, middle-aged Augustinian monk, Gregor Mendel, rose solemnly before the Natural History Society in the town of Brünn, in what is now Austria, and read a scientific paper which enumerated the basic laws of heredity. For over two thousand years these secrets had eluded the greatest thinkers humanity had to offer, but the reaction to "Experiments in Plant Hybridization" was hardly dramatic. His audience could hardly wait for Mendel to finish reading his paper before turning their attention to other matters. Mendel and his work were quickly, quietly and completely ignored. The scientific community was far too busy debating Darwin's theory of evolution to pay any but scant attention to the seemingly amateurish and presumptuous experiments of an obscure Austrian monk.

No one believed that Mendel's experiments were anything more than a strange hobby of a strange churchman. It is suspected that his fellow monks elected Mendel abbot of the monastery in 1868 in the hope that his new duties might take up enough of his time to put an end to his fiddling around in the garden. A few years later, a young American employed by a French botanical supply company visited Mendel at the monastery and asked him about his pea plant experiments. Mendel reacted by trying to change the topic. Apparently even he shared the view that his research was pure foolishness. It wasn't until the turn of the century, well after his death in 1884, that Mendel's work was rediscovered and he was acknowledged as the Father of Genetics.

Gregor Mendel was an unlikely candidate for such an achievement. Ironically the man who would discover the basic laws of genetics was a man who could never pass a qualifying examination as a science teacher. The son of a farmer, Mendel was born in 1822. He spent much of his youth working in the family orchards and gardens. His interest in botany continued after he entered the monastery as a novice at age twenty-one. His superiors hoped to take advantage of his apparent aptitude in the biological sciences and sent the young churchman to the university in Vienna with the expectation that he would return to Brünn a science teacher. But his academic achievements were disappointing. He never qualified as a teacher.

Despite academic failure, Mendel's keen interest in botany would not subside. In 1857 he began eight years of historic experiments with garden peas. Mendel carefully designed his experiments to avoid the pitfalls that had beset previous researchers. The usual approach was to look at the overall appearance of a plant or animal and determine which parent it resembled most and figure out why. Results of such studies were often bewildering and self-contradictory. The problem was that they focused on a total array of characteristics. Many of the characteristics they looked at ranged over a broad spectrum of variations. Leaf color for example might appear as shades of green, rather than as separate colors altogether. Differentiating between shades of green required a very subjective judgment which contributed to the confusing results. Mendel decided he would use pea plants because they had characteristics which took distinctive forms—no confusion would be possible. He decided to study only those traits which lent themselves to clear-cut systematic observation and statistical analysis.

After several years of preliminary research, Mendel settled on seven specific characteristics in which there was always clear and consistent contrast. Some pea plants were tall, reaching 6 or 7 feet in height, while others were short, perhaps 9 to 18 inches in height. The difference between them was unmistakable. No one could ever confuse a short plant with a tall plant. Some pea plants had seed leaves (cotyledons) that were yellow, others green. Again there could be no problem distinguishing between the alternate forms this characteristic might take. Mendel, having finally decided upon seven characteristics for his experiments (Figure 4) disregarded all other characteristics of the pea plant. That was the key to his success.

The pea plant was perfectly constructed for Mendel's experiments. The reproductive organs were surrounded by the petals and usually matured before the flower opened. This meant that self-fertilization usually occurred, and each variety tended to be pure-breeding. Mendel raised several generations just to make sure his experimental plants really were pure-breeding. He wanted to be sure that his six-foot plants always produced offspring that were six feet tall, that his green-seed-leaf plants always produced offspring with green seed leaves, and so on down the line.

Prevailing opinion in Mendel's day held that heredity was a blending process, something like mixing blue and yellow paint together to get green. Once mixed, the original traits were lost forever; they could never be recovered intact. With each future generation they would be mixed and diluted again and again. Mendel proceeded to use his pure-bred

TRAIT	ALTERNATE FORMS	
SEED SHAPE	smooth wrinkled	Seeds were smooth and round or angular and wrinkled.
COTYLEDON COLOR	yellow green	Seed leaves were yellow or green.
SEED-COAT COLOR	gray white	Seed coat was white or gray.
POD SHAPE	constricted unconstricted	Ripe pods were constricted or not constricted.
POD COLOR	green yellow	Unripe pods were yellow or green.
POD POSITION	along stem top of stem	Flowers and pods were located all along the stem or only at the top of the stem.
STEM LENGTH	tall short	Pea plants were tall or short.

Fig. 4 After several years of research Mendel settled on seven characteristics which displayed clearly contrasting forms

plants to test this blending theory against reality. Carefully he brushed pollen from a six-foot plant onto the reproductive organs of a one-foot plant. He rubbed the pollen of a plant with yellow seed leaves on the reproductive organs of plants that always displayed the contrasting green-seed-leaf trait. He called the offspring of this cross-fertilization of two purebreds a hybrid, the term we still use today.

If the blending theory were correct, the offspring of a tall-short mating should have been intermediate in height—

perhaps three or four feet. However, Mendel's results completely contradicted the blending theory. When tall plants were crossed with short plants, the offspring were always tall. No matter how many times he repeated the experiment the result was always the same. Cross a six-foot pea plant with a one-foot pea plant, and you always got a six-foot plant. Similar results were obtained for each of the seven characteristics. (Figure 5)

What did these findings mean? Mendel reasoned that there were two "determiners" or "factors" responsible for the characteristics he studied. For height, there was a determiner that caused tallness and a determiner responsible for shortness. (Today these Mendelian "factors" are called *genes*, and although he never used the term, to avoid confusion we will use the word gene from this point on.) In the creation of the hybrid generation, each offspring received one gene from each parent, and each gene was different. Tall plants contributed a gene for tallness and short plants a gene for shortness. It was obvious that these genes did not have equal power in influencing the physical appearance of the hybrid plant. The gene that expressed itself in the hybrid offspring, Mendel called *dominant*, the other *recessive*. The gene for tallness was dominant over the gene for shortness. The dominant and recessive genes in the pea plant experiment can be summarized in a chart. (Figure 6)

These empirical observations served as the basis of the Mendelian Law of Dominance: When a cross occurs between organisms pure for contrasting traits, only one appears in the hybrid offspring.

DOMINANT AND RECESSIVE GENES

What happens in the hybrid cross can be seen in a diagram. (Figure 7) The dominant gene is symbolized by a capital

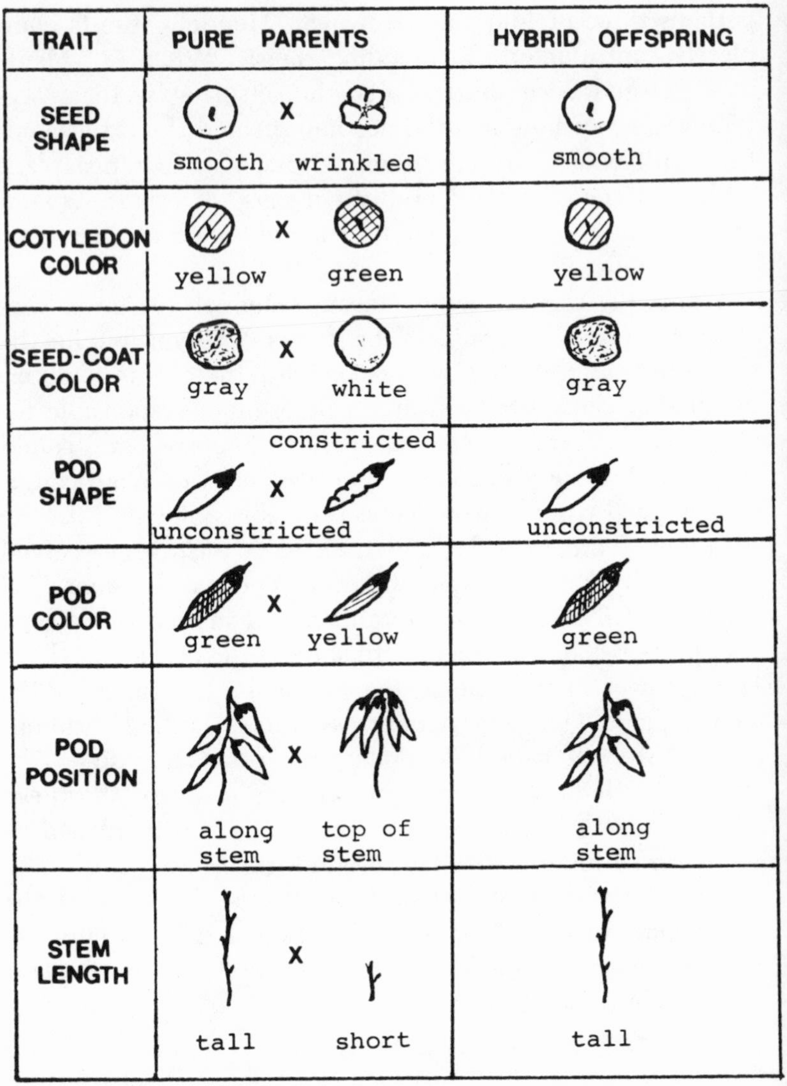

TRAIT	PURE PARENTS	HYBRID OFFSPRING
SEED SHAPE	smooth x wrinkled	smooth
COTYLEDON COLOR	yellow x green	yellow
SEED-COAT COLOR	gray x white	gray
POD SHAPE	unconstricted x constricted	unconstricted
POD COLOR	green x yellow	green
POD POSITION	along stem x top of stem	along stem
STEM LENGTH	tall x short	tall

Fig. 5 Results of the hybrid cross

CHARACTERISTIC	DOMINANT TRAIT	RECESSIVE TRAIT
SEED SHAPE	smooth seed	wrinkled seed
COTYLEDON COLOR	yellow seed leaf	green seed leaf
SEED-COAT COLOR	gray seed-coat	white seed-coat
POD SHAPE	unconstricted pod	constricted pod
POD COLOR	green pod	yellow pod
POD POSITION	along stem	top of stem
STEM LENGTH	tall	short

Fig. 6 Summary of dominant and recessive traits in pea plants

letter and the recessive gene by the corresponding small letter. For convenience we can say that **T** stands for the tall gene, and **t** for the short gene. Mendel believed each plant had two genes for each characteristic, one contributed by each parent. If this were true, plants that were pure for tallness would have to have a **TT** genetic makeup. Those pure for shortness would be **tt**. The diagram shows the genetic possibilities of the hybrid cross. The letters on the outer perimeter of the large square represent the gametes produced by the pure-bred parents. The boxes within the large square illustrate the possible genetic makeup of an offspring. The hybrid receives one gene from each parent. From the tall parent it always receives a **T**, from the short parent a **t**. All of the offspring would have a **Tt** genetic makeup. Since **T** dominates over **t**, the offspring all appear as tall.

Mendel's hybrids with their **Tt** genetic makeup for height and his **TT** pure-breds displayed the same tall physical appearance. This brings us to two important concepts in genetics: the *genotype* and the *phenotype*. Every organism has a genotype which refers to the unique combination of genetic information inherited from the parents. The phenotype refers to the physical appearance of an organism. The **TT** and **Tt** plants had different genotypes but exhibited identical phenotypes. We will discuss these concepts again in later chapters.

Once he understood what happened in the hybrid generation, Mendel wondered what had happened to the recessive gene. Had it disappeared forever? Was it destroyed or altered in some way? To find the answer, Mendel decided to plant the seeds produced by the self-fertilization of the hybrid plants.

The results of this planting were that for each and every one of the seven characteristics, approximately one-fourth of the offspring showed the recessive trait that had seemed to disappear in the hybrid generation. (Figure 8) One-fourth of these plants were short, totally unlike their hybrid tall (**Tt**) parents, but exactly identical to their **tt** grandparents. The other three-quarters resembled the hybrid generation in height.

Clearly, the recessive gene had not been changed or destroyed. Its influence had merely been masked by the dominant gene in the hybrid generation, but it had been carried within the hybrid and passed on to the next generation absolutely unchanged. The appearance of the recessive trait in a one to four ratio is not magic or mysterious. You can see why by using the same type of square diagram we used earlier. (Figure 9)

This square represents the probabilities of random recombination of genes in the offspring. Each hybrid can produce two types of male gametes (**T** and **t**) and two types of female gametes (**T** and **t**). How these genes recombine in the formation of the next generation is totally random. The **T** of a male parent has just as much chance of uniting with a **T** or **t** of the female, so the possible results are **TT** or **Tt**, as shown in the first line of the diagram. The **t** of that same

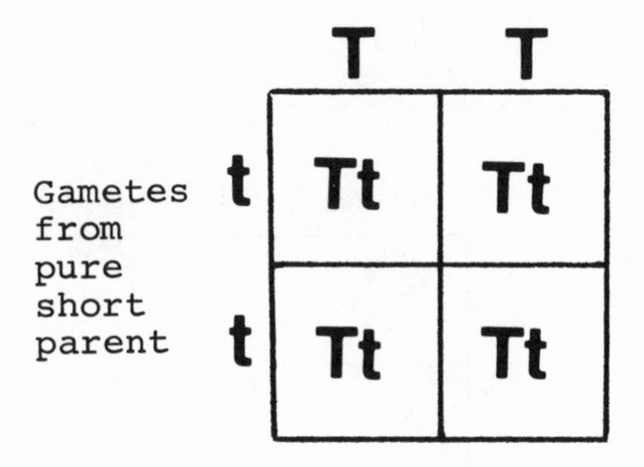

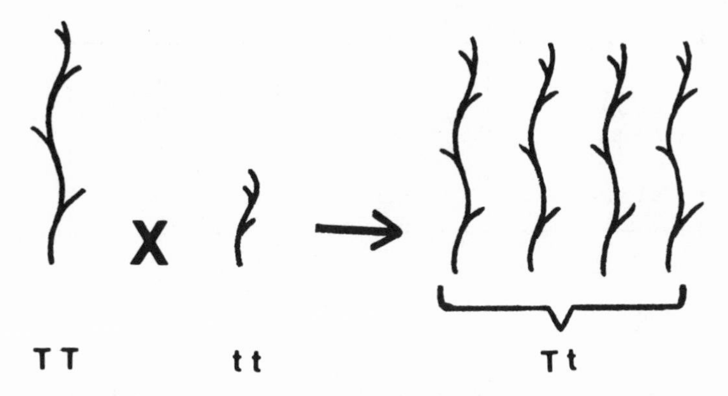

Fig. 7 When plants pure for contrasting traits are crossed, the offspring exhibit only the dominant form of the characteristic

male, likewise, has the same chance of uniting with **T** or **t** of the female. The possible results, **Tt** or **tt**, are illustrated in the second level of the diagram. Count up the boxes and the results are clear: one-fourth **TT**; two-fourths **Tt**; one-fourth

TRAIT	HYBRID SHOWS	NO. SHOWING DOMINANT	NO. SHOWING RECESSIVE	RATIO
SEED SHAPE	smooth	5,474 smooth	1,850 wrinkled	2.96:1
COTYLEDON COLOR	yellow	6,022 yellow	2,001 green	3.01:1
SEED-COAT COLOR	gray	gray	224 white	3.15:1
POD SHAPE	unconstricted	882 unconstricted	299 constricted	2.95:1
POD COLOR	green	428 green	152 yellow	2.82:1
POD POSITION	along stem	651 along stem	207 top of stem	3.14:1
STEM LENGTH	tall	787 tall	277 short	2.84:1

Fig. 8 Results of the self-fertilization of the hybrid plants

tt. Since **TT** and **Tt** are phenotypically the same, three-quarters of the plants are tall. Since there is no dominant gene present in the **tt** plant to mask the shortness trait, the gene for shortness expresses itself. Notice that the hybrid tall

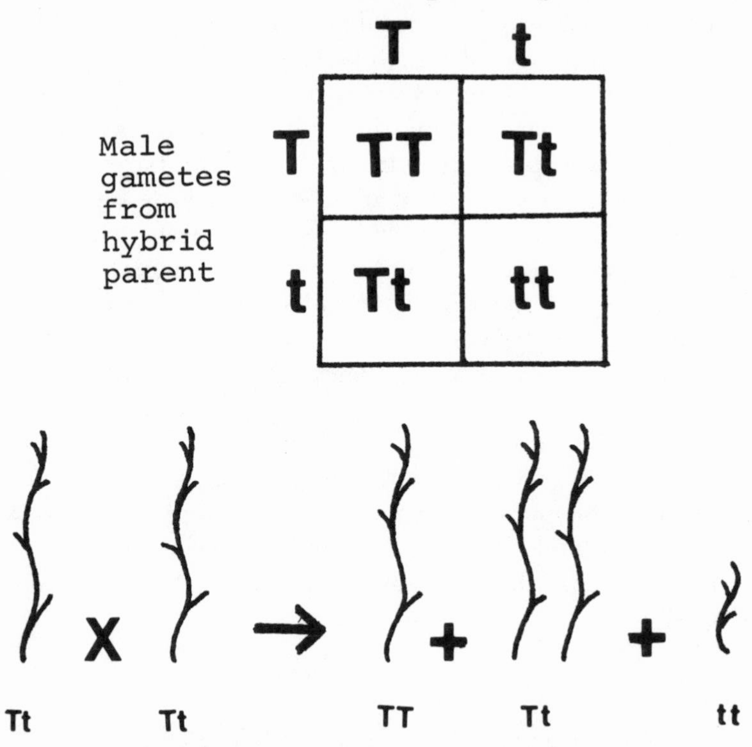

Fig. 9 The recessive trait may reappear in definite percentages in the offspring of the hybrid generation

parents produced offspring that were pure for shortness!

These observations led to the Mendelian Law of Segregation: Genes in the same pair are passed on separately when an organism forms gametes and reproduces. The recessive trait which disappears in the hybrid generation may reappear in future generations in definite percentages.

Mendel next turned his attention to what happens when more than one characteristic is studied. Are different characteristics inherited together or separately? Is there some kind of interaction between characteristics? Does the inheritance of one influence the inheritance of another? What happens when plants pure for tallness and green pods (both dominant) are crossed with plants pure for shortness and yellow pods (both recessive)? When Mendel performed these crosses he again found that only the dominant trait appeared in the hybrids. It didn't matter which parent contributed which genes. When he crossed plants pure for tallness (dominant) and yellow pods (recessive) with plants pure for short (recessive) and green pods (dominant), the offspring displayed only dominant traits—tall with green pods. (Figure 10) Recessive traits again reappeared in the next generation in definite percentages when the hybrids were permitted to self-fertilize.

The seven characteristics Mendel studied were inherited independently of each other. Any possible combination of traits was due purely to chance. These empirical observations led to Mendel's Law of Independent Assortment: Every characteristic is inherited independently of every other characteristic.

MENDEL WITH A GRAIN OF SALT

Mendel's laws are simplifications of a complex reality. There are numerous exceptions. If taken literally, they can lead to an erroneous idea of how heredity works. There are a number of pitfalls to be aware of.

One of these pitfalls is demonstrated by this quotation from a *New York Times* article in 1970: ". . . each individual carries two genes, or bits of genetic instruction for every human characteristic: one from each parent." There is serious error here. True, the characteristics Mendel focused on were apparently controlled by a single pair of genes in each

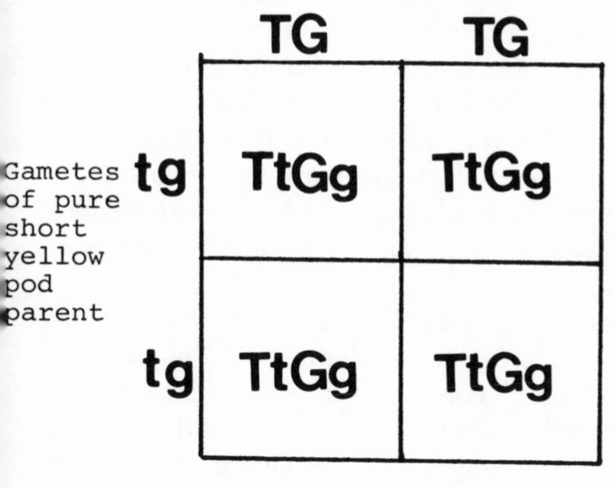

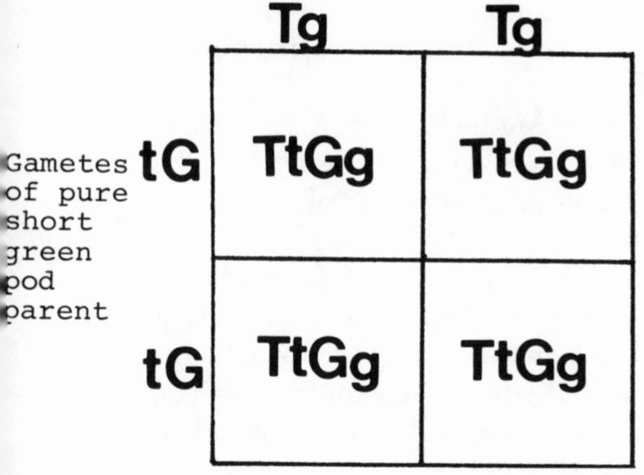

Fig. 10 In a double hybrid cross, no matter which parent contributes which genes, only the dominant trait is expressed

case, which displayed contrasting forms in the phenotype. But it is incorrect to conclude from this, as the *Times'* article did, that all characteristics are the result of the action of a single pair of genes. Most characteristics are not controlled by a single pair of genes but are in fact *polygenic*. What form they take is the result of the mutual interaction of numerous gene pairs. In many cases there are more than two possible forms a gene may take. Each individual may indeed have two genes per pair, but the genes may take a variety of forms.

It is also a gross oversimplification to say that traits always appear in contrasting forms. Remember, it took Mendel years of exhaustive research finally to find seven characteristics that did manifest themselves in distinctive forms. He decided to ignore characteristics which were not sharply contrasting, and it wasn't easy, because most characteristics vary only in degree. People, for example, are not simply tall or short. Human height varies over a continuum, and it is frequently very difficult to locate a dividing line that is not subjective. Six feet may be tall, or it may be short, depending upon your point of reference.

Another exception to Mendelian law is the phenomenon of *intermediate dominance*, in which offspring resemble neither parent, but exhibit a different form of the characteristic. Among certain breeds, a red bull crossed with a white cow will give rise to calves that are roan (reddish-brown with gray intermixed) in color. And among Andalusian fowl a cross between a black and a white parent results in offspring that appear blue. Had Mendel conducted his research on organisms such as these, he might have concluded that the blending theory was correct.

One should also keep in mind that a given genotype does not necessarily guarantee a specific phenotype. An individual inherits genes that give the potential for developing certain traits under the right conditions. The influence of the environment on the genotype is critical. A pea plant with a genotype giving it the potential to be tall would likely vary from the norm if it were affected by negative environmental influences such as poor soil, insufficient water, inadequate sunlight, or adverse climatic conditions. The genotype gives the potential to develop a specific phenotype.

Mendel's Law of Independent Assortment cannot be taken literally, either. Today we know that genes are located on chromosomes in the cell nucleus. Individual genes do not float around randomly. Each has a designated location, or locus, on a specific chromosome. In general, there is a likelihood that genes on a chromosome will be inherited together as a group. Thus, the idea of independent assortment more accurately applies to genes that are on different chromosomes.

Mendel was a lucky man. The traits he studied occurred in easily distinguishable forms. They were controlled by genes located on different chromosomes and were therefore inherited independently of each other. If two or three of these genes had shared the same chromosome, they would have been inherited together. Mendel might never have made sense out of his observations. If the pea plants he chose had not been pure-breeding, he might have run around in circles forever.

As a matter of fact, Mendel's luck did run out when he tried to duplicate his experiments with hawkweed plants and honeybees. These organisms didn't breed in the same reliable way as the pea plants. He failed to reproduce his findings.

Mendel's work has to be taken with a grain of salt. Even the earliest geneticists recognized that Mendel's laws were simplified models of a complicated reality. But it would be foolish to discount his contributions to genetics. We can't blame Mendel if inaccurate applications or interpretations of his work have persisted. What cannot be denied is that Mendel was the first person in human history to formulate a coherent statement of the following points:

——Genes (or determiners or factors as he called them) occur in pairs.

——Genes are passed on unchanged from generation to generation.

——Some genes are dominant and others recessive.

——Each parent contributes one gene for each pair to the offspring.

——A recessive gene which seemingly disappears in one generation may reappear in future generations.

——The members of each gene pair segregate when gametes are formed.

——Some kind of independent assortment takes place, contributing to variation in the population.

These were earthshaking contributions to the understanding of heredity—a long way from blood and vapor theories, homunculi, spores and Chinese boxes. Incredibly, Mendel's work was ignored by his contemporaries. Nearly four decades would elapse before Mendel's work would be rediscovered and appreciated in the scientific world.

CHAPTER THREE

The Cell: The Basic Unit of Life

Robert Hooke (1635–1703) sliced off a piece of cork one day in 1665, placed it under his microscope, and noticed that it contained a honeycomb network of empty, rectangular chambers, which reminded him of cells, the small rooms of monks in monasteries. Through the years to follow, other examples of cells were discovered by scientists throughout Europe. Unlike cork cells, which are the no-longer-living cells of cork oak tree bark, living cells were not empty, but filled with a jellylike substance that came to be called protoplasm. Biologists began to suspect that cells could be found in all living tissue.

By 1838, M.J. Schleiden pulled together much of the research and speculation, and presented a theory that held that all plants were constructed of cells. A year later, another German scientist, T. Schwann, pointed out that animals, too, were composed of cells. Together they share the credit for formulating the cell theory.

Once it was established that cells themselves are living

things, and that all animals and plants are built from combinations of cells arranged in accordance with rules determined by nature, the accurate understanding of how the cell lives and reproduces finally became possible.

Over the years there have been many improvements in the cell theory. The early understanding that cells were composed of protoplasm has been superseded by a more sophisticated view that the tiny cell itself is a complex organization of even tinier *molecules* (the smallest unit into which a substance can be divided and still retain its original character) and substructures. This complex organization within the cell is closely intertwined with genetics, so it would be wise to take a closer look at the structure of cells.

CELL MEMBRANE

All cells are enclosed by a cell membrane, which is a living part of the cell. It provides shape, support and protection for the rest of the cell. In addition, it permits nutrients, wastes and water to pass in and out of the cell as necessary. Cell membranes are "semi-permeable," permitting some substances to pass into the cell, but not others. What determines what is allowed to enter the cell and what is not? Apparently, the size of the molecules is not a factor. Amino acids, for example, have large molecules but are permitted to pass through the membrane. Other chemicals with much smaller molecules are blocked from entering the cell. Evidently the membrane somehow possesses an ability to determine what material the cell needs and what is either harmful or unnecessary. A selective pumping action forces through only those substances required for life-sustaining functions, including amino acids.

CYTOPLASM: THE FACTORY OF THE CELL

When you examine a cell under the microscope, you can see that the protoplasm, the substance of the cell, is divided into two distinct parts. Toward the center of the cell the protoplasm appears darker and denser. This is the nucleus. The less dense material surrounding the nucleus is called the cytoplasm. (Figure 11) Within the cytoplasm is a network of substructures carrying out specialized tasks necessary for the maintenance of life. So much work goes on in the cytoplasm that it is frequently called the factory of the cell.

Tubelike membranes within the cytoplasm (endoplasmic reticulum) provide a network of pathways through which materials may be transported from the cell nucleus to the cytoplasm. In the walls of the endoplasmic reticulum are tiny bodies called ribosomes, which are rich in the chemical RNA (ribonucleic acid) and play a crucial role in heredity.

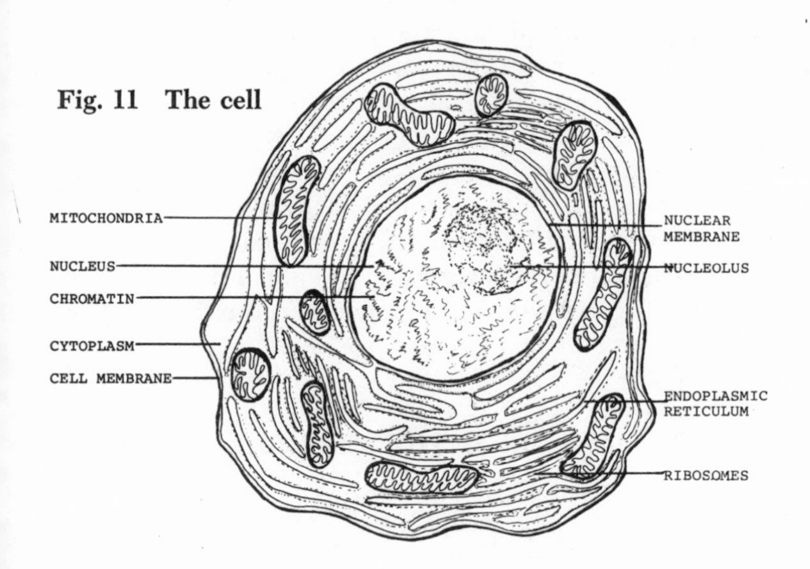

Fig. 11 The cell

Another important set of substructures in the cytoplasm are the mitochondria—the powerhouses of the cell. Nutrients such as carbohydrates, fats and proteins which have been passed into the cell through the membrane are collected and burned at the mitochondria to release energy, carbon dioxide and water. The energy produced is manufactured into energy-storing molecules that can be retained until needed by the cell. At the right moment, special chemicals called enzymes, produced at still other sites in the cytoplasm, initiate reactions to release the energy so the cell can do its work.

Some cells require more energy than others, depending on the special job each has to perform. The number and placement of the mitochondria powerhouses are related to the energy needs of the particular cell. Muscle cells require much energy. Here the mitochondria are located in rows near the muscle fibers so they can have the maximum effect when releasing their energy-bearing molecules. In sperm cells, the mitochondria are found at the base of the long, muscular tail which provides propulsion.

There are numerous other substructures in the cytoplasm responsible for producing specific enzymes and other substances needed for the maintenance of life. There is no need to describe any more of them here; by now you get the idea. The cell is not just a tiny droplet of living material. It is a microscopic computerized factory, with countless substructures, or departments, executing definite tasks and precise functions, at exactly the right moment, in the right way, in a smooth and remarkably efficient manner.

THE NUCLEUS: THE CENTRAL CONTROL

Like all computers, the cell has a central control to supervise operations. This role is played by the nucleus, which regulates the activity and production of the entire cell. And like all computers, the central control requires a program, or set of instructions, that tells it what to do under what conditions and at what time. This information may be written or coded in a variety of ways, filed away, stored and recalled for consultation when necessary. It can also be reproduced for use in even newer computers. This is as true for the IBM computer as it is for the cell. Man-made computers store their programs on magnetic tapes or punch cards. In the cell, the precise instructions for the maintenance and continuation of life are coded in molecules of DNA.

This program of instructions is called the *genetic code*. Under the right environmental conditions it provides for the operation, development and reproduction of individual cells as well as the entire complex organism. This natural computer program is far superior to even the most sophisticated computer program devised by humans. For nature, the interference of a "programmer" to correct or update information is impossible. Nature has developed a self-contained system with built-in feedback that signals the next phase in the life cycle to begin and that contains within it an ability to change and correct itself. The nucleus controls the activity of the cell, but the genetic program tells the nucleus what to do.

It is the information in the genetic code that enables the cell membrane to differentiate between substances that are needed and those that are harmful. It is the genetic code that is ultimately responsible for the mitochondria being in the right location in the muscle cell, for the production of the right enzymes to release the stored-up energy, and so on. It is this genetic information that tells the cell whether it will develop as a brain cell or a liver cell. There is much more to genetics than figuring out where your blond hair comes from. The genes are responsible for the integration of all aspects of life.

The genes are found in the nucleus in long chains called chromosomes. Individual genes are so tiny, so complicated in form, that no one has ever seen one. Chromosomes, however, are larger and susceptible to detailed study. With the sound basis of the cell theory, and advances in technical methodology, investigators were able to concentrate on the subcellular and molecular levels of life, which in the end shed more light on how the larger system worked. It was by concentrating on the smaller, that the larger picture became clear and focused.

CHAPTER FOUR

Chromosomes: The Carriers of Heredity

When the cellular research pioneers began studying the concentration of protoplasm in the center region of the cell, they didn't really know what they were looking at. The cell was the basic unit of life; every cell had a nucleus—so they figured it must be important. Exactly how important and what role it played in the life of the cell was still a mystery. To accumulate the necessary information that would someday lead to an answer, they began by compiling detailed diagrams and descriptions of their observations.

Certain privileges go along with being a pioneer in a new field, and one of these privileges is the right to pin names on the new and unknown things you find. One of these things they had to name was a strange mass of granules and threads present in the nucleus. Frequently biologists use a technique called *staining* to make it easier to study things under the microscope. A chemical dye is applied to the material which improves contrast and clarity when the slide is examined. (Perhaps you remember staining onion cells in science class.) The mystery substance readily accepted the dye coloring. Scientists gave it a name derived from a Greek word meaning affinity for color: chromatin.

Chromatin was not always visible. As early as 1848, it was clear that when cells were in the process of dividing—splitting in half to create two new cells—the chromatin was replaced by tiny, rodlike structures, which also accepted chemical stain. Because they appeared during cell reproduction, it was generally assumed that these rods, which were named chromosomes, were important components in the hereditary process.

Attention focused on chromosomes and the data began to pour in over the next several decades. Every cell had chromosomes. All individuals in the same species had the same number of chromosomes. Within an individual, every cell had the same number of chromosomes. This raised an intriguing dilemma. How could nature keep the number of chromosomes in a species constant generation after generation if a new organism was produced by the fusion of reproductive cells from two parents? If the parents had the same number of chromosomes as each other, why didn't their offspring have double the number of chromosomes in its cells?

Fig. 12 The chromosome dilemma. If each gamete had 46 chromosomes, the number of chromosomes would double each generation

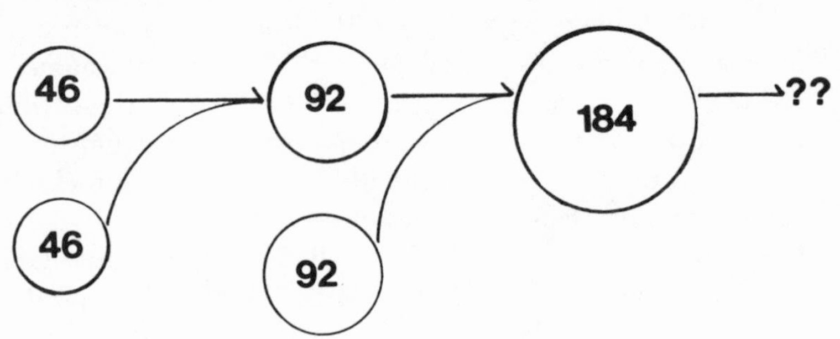

Why didn't the number of chromosomes double with each generation? (Figure 12)

August Weismann (1834–1914) made the first significant attempt to solve this puzzle. Nature, he said, must have two different types of cell division at its disposal. The first would account for normal body growth, producing new cells with the same genetic material. The second method would be responsible for the production of gametes, the reproductive cells, with only half the full genetic complement. Thus, when egg and sperm united, they were restoring the normal number of chromosomes, not doubling it.

The main thrust of Weismann's speculation was correct. He was mistaken only in the details. Normal body growth is indeed attributable to a cell division, called *mitosis*, which produces daughter cells genetically identical to the mother cell. Today, we know that a chromosome is a long chain of genetic material. As the cell prepares to divide, each segment of genetic material, or gene, produces a replica of itself. In this manner, an identical chromosome is manufactured, gene by gene, within the nucleus. When the cell actually divides, these identical chromosomes pull apart, each headed for the nucleus of one of the new daughter cells, guaranteeing each daughter cell a full-scale version of the same genetic material that was present in the mother cell. This mitotic division accounts for all normal growth from the moment of fertilization.

Weismann correctly saw that the only way to avoid giving the offspring a double dose of hereditary information was through a cell division that reduced the amount of genetic material in the gametes by one-half. The term he coined for this process, *reduction division*, is still widely used today, though the scientific name for it is *meiosis*. Today, we know that each species not only has a fixed number of chromosomes, but that these chromosomes generally come in pairs.

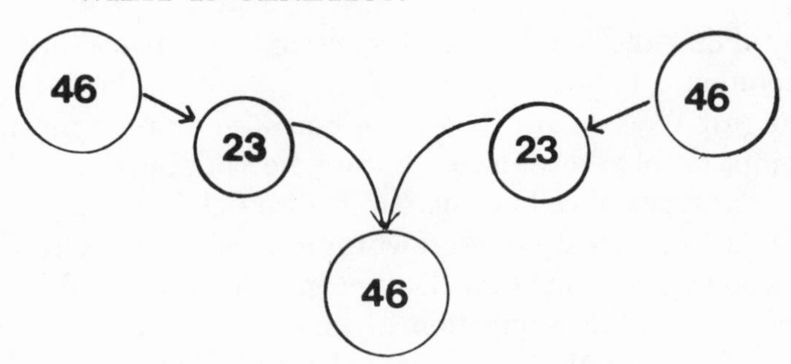

Fig. 13 Meiosis reduces the number of chromosomes in the gametes by one half, so that when fertilization occurs the normal number of chromosomes is restored

Humans, for instance, have 23 pairs, or 46 chromosomes. Meiosis makes sure that each gamete carries only 23 chromosomes, one from each pair. (Figures 13 and 14)

CHROMOSOMES AND GENES

Genetic material assumes different forms at various stages in the life cycle of the cell. Sometimes it appears as diffuse chromatin threads, sometimes as squat chromosomes. As the cell prepares for division, the genetic material passes through a series of alterations, concentrating and contracting itself until it takes the shape of chromosomes. Nature is too thorough for such changes to be random or purposeless. The contraction of chromatin into chromosomes helps make sure that the separation of genetic information during cell division will be accurate. In mitosis all hereditary information within the mother cell nucleus is passed on to the daughter cells, which, in turn, pass the same information on to their daughter cells. The average adult human body may utilize as

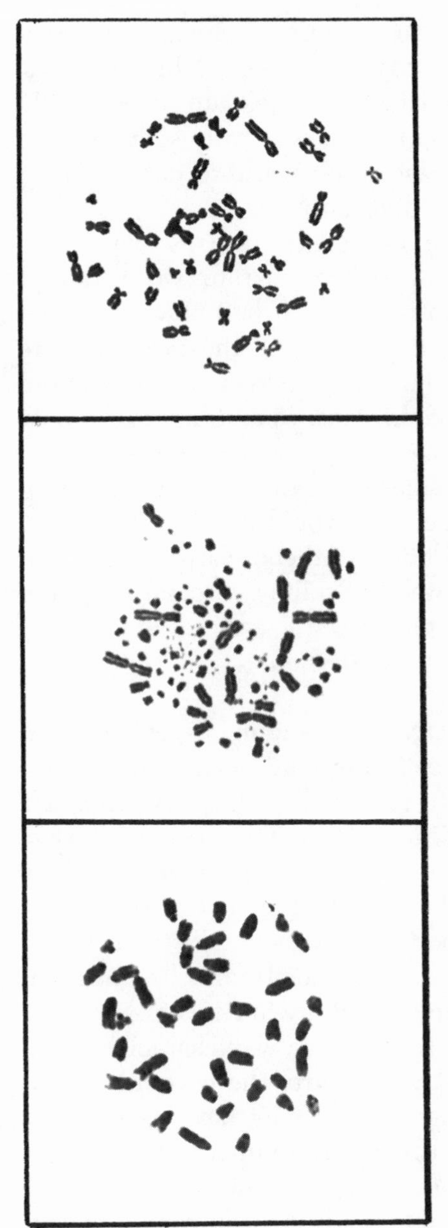

Fig. 14 Human (top), chicken (middle), mouse (bottom), chromosomes. Each species has its own characteristic set of chromosomes. These vary in number and shape from species to species (Courtesy of the Cytogenetics Laboratory, New York University Medical Center)

many as a 100 billion cells in a lifetime, which means that mitotic cell division would have occurred 50 billion times. The repetition of such an intricate and delicate process without significant mishap 50 billion times is truly something to marvel at. The contraction of chromatin into chromosomes helps assure an accurate cell division.

How? Imagine that the chromatin threads were long strands of yarn. It would be easier to separate two strands of yarn if they were rolled up into a tight ball of yarn than if they were tangled up together. Imagine the mess if there were 46 strands of yarn tangled up together. If you tried to separate them under these conditions, you might accidentally rip them or break them, or even tie them in knots. Now if your yarn were really chromosomes, things would be even more confusing because each one would have duplicated itself, so temporarily there'd be 92 of them. Nature has developed its own automatic method of rolling chromatin up into tight little chromosomes to minimize the dangers of tangling, ripping and breakage during cell division.

A gene is a segment of DNA. It contains a specific coded message that instructs the cell to produce some kind of protein which affects a characteristic of the cell or organism as a whole under the right conditions. Perhaps the coded message signals a cell substructure to produce an enzyme to control the burning of fuels and the release of energy at the appropriate moment. Perhaps the DNA message tells the cell that in response to certain stimuli or signals from cells surrounding it, or other external sources, that it will develop as part of a muscle and how to organize itself in a way characteristic of muscle cells. Or perhaps it will control the color of the organism's eyes or hair. A human cell may have as many as 200,000 genes in a single cell, each with important jobs to do, and many handling a number of responsibilities.

Nature has taken the trouble to assign each gene a specific

locus on a specific chromosome to make certain that none of them get lost in the shuffle during cell reproduction. It's simply a question of efficiency and accuracy. If genes were not at designated spots on the chromosomes, the nucleus would be anarchy when the genes replicated. Two hundred thousand genes would become 400,000, all floating around randomly in a puddle of protoplasm. It would be incredibly difficult and time-consuming to sort out the genes to make certain that each daughter cell received its complete set of genetic material. It's much simpler to deal out the tightly packed chromosomes.

CHROMOSOME VISIBILITY

A geneticist sees a lot more when studying a chromosome today than Weismann ever did. Microscopes are more powerful so that greater detail in structure, behavior and arrangement is observable today. Improvements and refinements in staining procedures help differentiate the sometimes subtle characteristics of chromosomes. Sophisticated cell culturing techniques help obtain cells in the process of dividing which therefore offer greater visibility of chromosomes. The very material placed on slides and examined in the laboratory is apt to be better suited for study than it was a hundred years ago. Special cameras fitted to microscopes provide photographs of chromosomes which can be enlarged even further. These advances create a situation where the study of chromosomes is daily more precise.

Since human chromosomes are distinguishable from each other, geneticists have been able to assign each pair a numerical designation. So routinized has the analysis of chromosomes become, that technicians can take a photograph of a human's chromosomes, cut them out, sort them and rear-

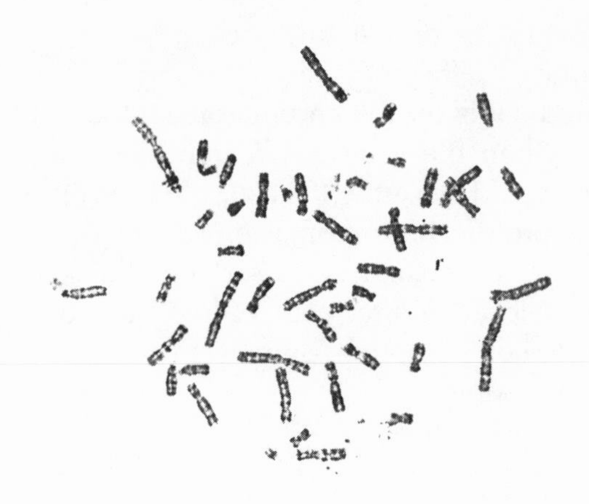

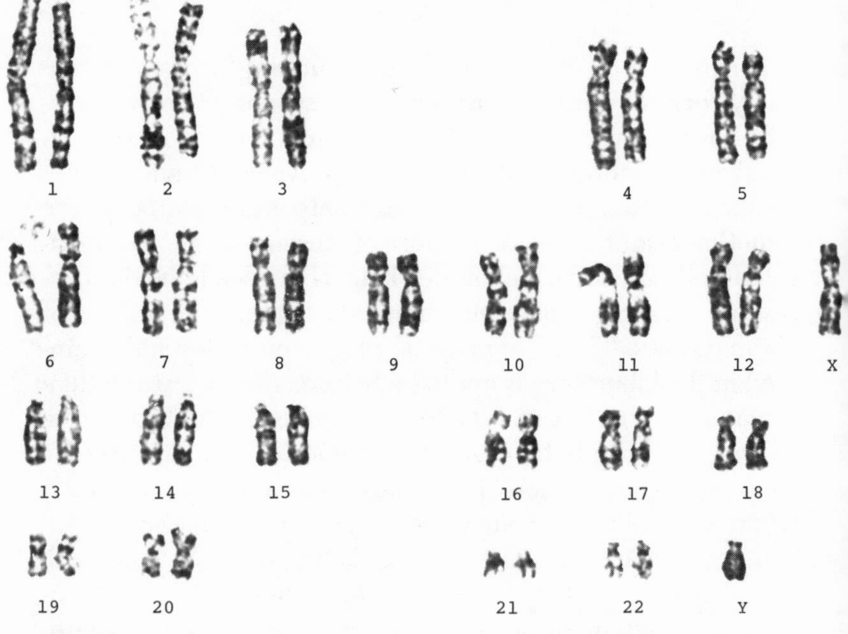

Fig. 15

range them pair by pair like a jigsaw puzzle. This process is called karyotyping, and is a widely used technique in checking for chromosomal abnormalities. (Figure 15)

CHROMOSOMES AND VARIABILITY

We have mentioned that there is a tendency for the inheritance of genes on the same chromosome to be linked together, and that the concept of independent assortment should more appropriately be applied to these *linkage groups* (genes sharing the same chromosome). Each chromosome is inherited independently of all other chromosomes; there is no linkage between chromosomes. Half your chromosomes come from your father's sperm and half from your mother's egg, but maternal and paternal chromosomes are in no way, shape or form linked together. When you produce

Fig. 15 Chromosomes of a human male, with karyotype below. Karyotypes are made in the following way. Cells are taken and cultured in a supportive growth medium so that an adequate number of dividing cells can be collected. A chemical which interrupts the division process is used to lock these cells in that phase of division in which the chromosomes are visible and individually identifiable. Then the cells are swollen with a dilute salt solution so that the chromosomes won't be tangled up. The fixed cells are then dropped on slides, treated, and stained to help make the internal structure of the chromosomes more visible. The cells are examined and photographed under the microscope at an enlargement of about 1250 times, and then the photographs themselves are enlarged from 4 to 10 times. Individual chromosomes are then cut and arranged in a standardized manner, which has been established by international agreement of geneticists throughout the world. This arrangement is called a karyotype (Courtesy of the Cytogenetics Laboratory, NYU Medical Center)

gametes, the maternal and paternal chromosomes in your cells will be rearranged completely at random. It is possible that all the chromosomes in a particular gamete of yours might have originated with your father, or 16 of them could be paternal and 7 maternal in origin, or 10 paternal and 13 maternal, etc. It is completely determined by the blind laws of chance. If a species has four pairs of chromosomes, sixteen arrangements are possible. (Figure 16)

This constant scrambling and reassortment of chromosomes and their genes is only one of the ways that nature promotes variation within a species. Another way is by *chromosome crossover*. Linkage groups are not permanent, once and forever; they are subject to change. We can say that linkage groups generally tend to be inherited together—but not always. Genes are not irrevocably or unalterably bound to their original chromosomes. Sometimes genetic material at the same locus on corresponding chromosomes swaps places during cell division. Corresponding genes apparently have a strong mutual attraction. Just before chromosome pairs pull apart during reduction division, for example, they experience a phenomenon called synapsis, in which they wrap tightly around each other, as if drawn together by a powerful magnetic force. This attraction is so powerful that sometimes a break occurs in the chromosomes as they pull apart before the production of the gametes and some genetic material may cross over from one chromosome to the other.

Chromosome crossover can be illustrated in diagrammatic form. (Figure 17) If we designate each gene in the chain with a letter, we can say that one of the duplicated chromosomes is **A-B-C-D-E-F-G-H-I** and the corresponding duplicated chromosome is **a-b-c-d-e-f-g-h-i**. During synapse these chromosomes wrap tightly around each other. A break occurs, some of the genes—let's say **F-G-H-I** and **f-g-h-i**—switch places. The gametes resulting from this reduction

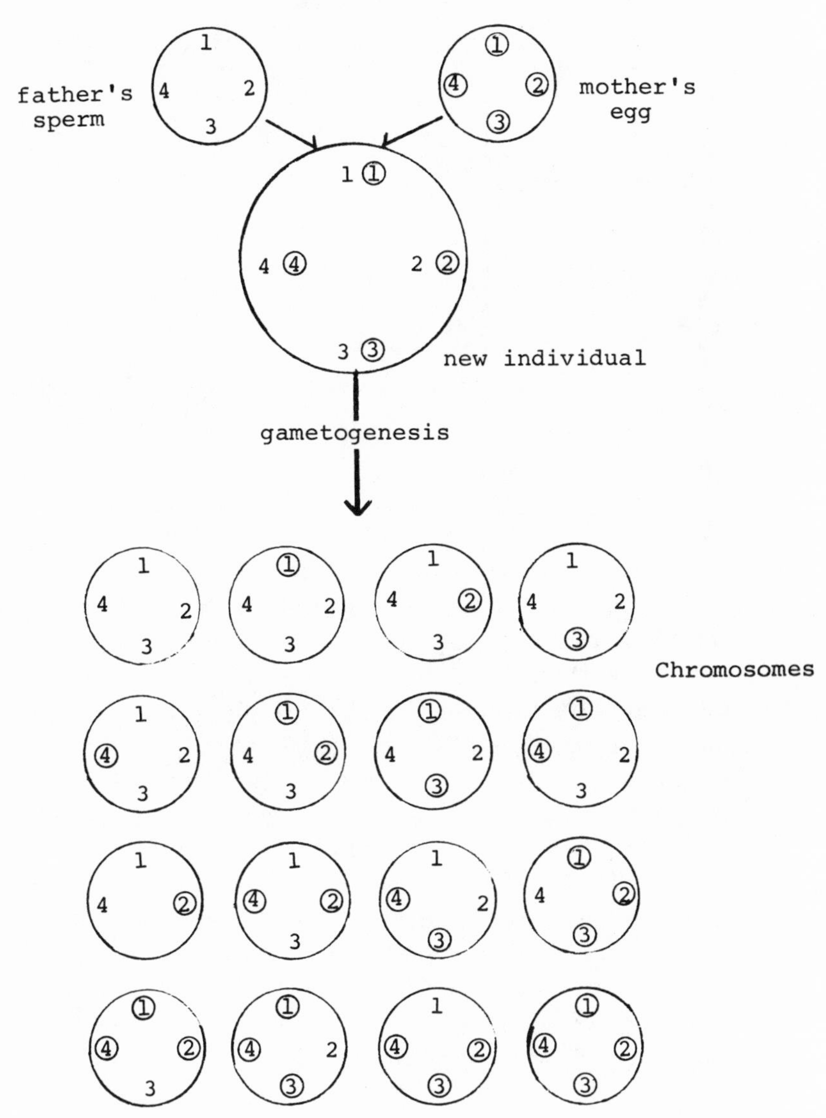

Fig. 16 Random assortment of four chromosomes producing sixteen possible combinations in the gametes

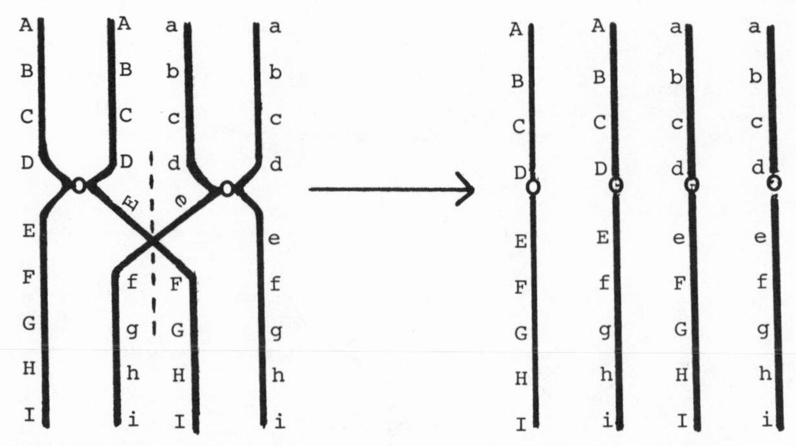

Fig. 17 Chromosome crossover during the production of gametes

division will differ from those we would normally expect. Only two would be identical to the original chromosomes in the mother cell; the other two would be a recombination of the genes, appearing as **A-B-C-D-E-f-g-h-i** and **a-b-c-d-e-F-G-H-I**. In the event that either of these chromosomes would be involved in fertilization, this new arrangement of genetic information would be passed on as a linkage group, promoting even more genetic variation than originally seemed possible. It's perfectly possible that a gene originally carried on a maternal chromosome will wind up on a paternal chromosome. Henceforth its inheritance would be linked with traits previously appearing on the paternal side of the family. Let's say for the sake of illustration—and this is certainly a fictionalized example—that the gene for blue eyes runs in your mother's family on the same chromosome as a gene for pointy ears. And let's say that crossover occurs. The gene for blue eyes may now be linked with a gene for rounded ears on your paternal chromosome, and even though blue eyes and pointy ears have been associated in your family for generations, you may have a child with blue eyes and rounded ears!

Crossover is not a rare event; it happens all the time. Two genes at extreme opposite ends of a large chromosome are almost as likely to be inherited independently of each other as they would be if they were on entirely different chromosomes.

In the introduction to this book we commented that variety is the essence of life. Nature has done her best to guarantee the widest possible variation in heredity, at the same time that it assures a certain uniformity. Humans have the capacity for broad-ranging variation, but babies are always little people. Nature has devised ways to take advantage of accidental errors that are bound to occur, like chromosome breakage, where the resulting crossover helps to strengthen variability. As the carriers of heredity, chromosomes simultaneously contribute to the stability and variability of the species.

CHAPTER FIVE

How Meiosis and Mitosis Work

Not too long ago doctors at a New York City hospital discovered a baby born with an extra set of chromosomes. Instead of chromosome pairs, there were chromosome triplets. The baby was dead at birth—there was no way it could survive. The old adage, "If two are good, three are better," does not apply when it comes to genetics. The cell's computer program is a precision instrument, forged through millions of years of evolution. The right amount, arrangement and balance of genetic information produce a healthy, properly functioning organism under the right conditions. Too much or too little of any ingredient is not good enough.

Extra genetic material is almost always harmful for humans. When mistakes happen—and they do—the baby born with extra chromosomes or extra pieces of chromosomes is usually abnormal. These birth defects may range from something inconsequential to something extremely serious. If genetic material is missing, the results are frequently even more drastic, producing harmful birth defects or death.

The maintenance of the chromosome number is critical for the smooth functioning of life.

MITOSIS

Mitosis is the cell division which maintains the chromosome number within a living thing after fertilization occurs. To do its job successfully mitosis has to satisfy three basic requirements:

1) it has to provide for exact duplication of the chromosomal material;

2) it has to provide a systematic method for evenly dividing this material;

3) it has to provide for the physical division of the mother cell into two independent daughter cells.

Textbooks usually divide mitosis into a number of phases on a rather arbitrary basis. But mitosis is a continuous process. Each phase merges almost imperceptibly into the next. For our purposes here, it isn't necessary to dwell on the specific characteristics of each phase, but only to give a general picture of how the process maintains the chromosome number and meets the three basic requirements mentioned above. (Figure 18)

Prior to cell division, the chromatin duplicates itself gene by gene. At the same time, the chromatin begins to contract to form the chromosomes and the nuclear membrane separating the nucleus from the cytoplasm disappears. The *nucleolus*, a substructure in the nucleus storing RNA (ribonucleic acid), also disappears. By the time chromosomes become visible, duplication has been completed. What we see under the microscope are actually duplicated, identical chromosomes, joined together at a single spot, called the *centromere*.

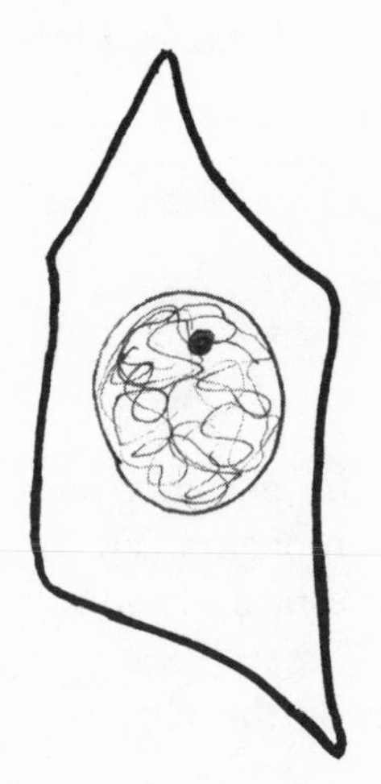

1. Chromosomes double in preparation for division

2. Doubled chromosomes contracted and identifiable

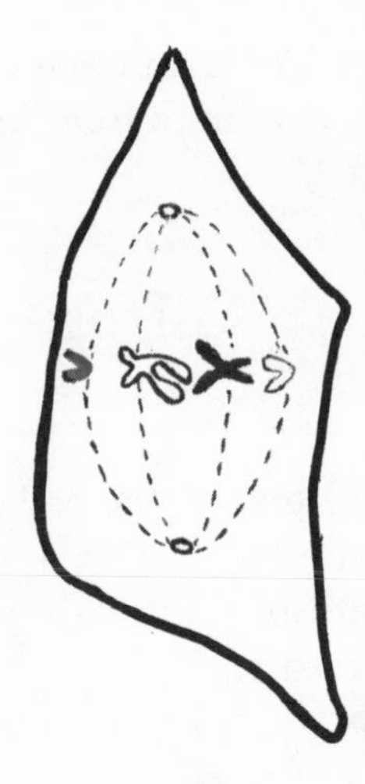

3. Doubled chromosomes line up randomly

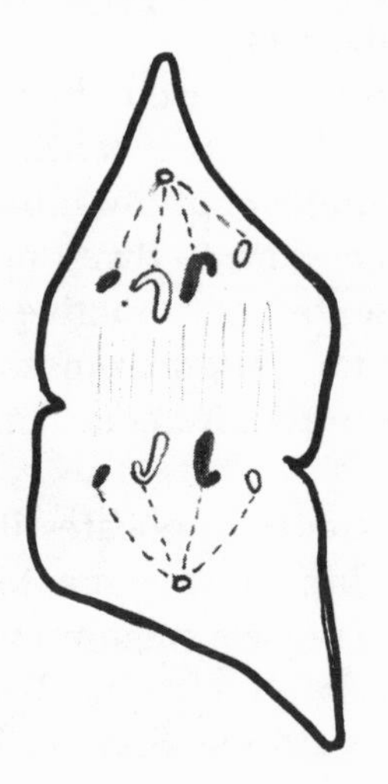

4. Centromeres divide and chromosomes separate

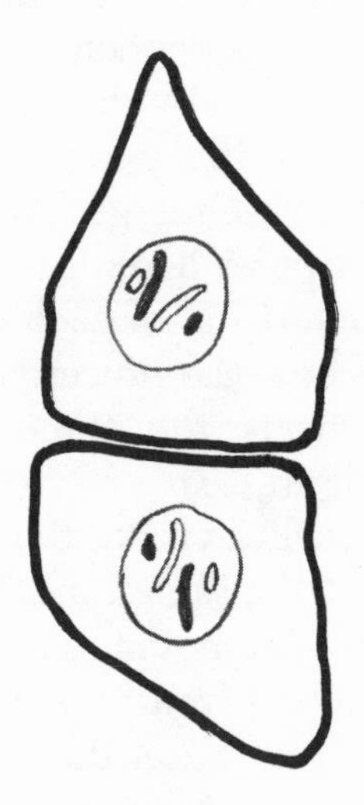

5. Equal daughter cells formed

Fig. 18 Mitosis

Systematic Division

The double chromosomes now move toward the center of the cell in a completely random manner. Corresponding chromosomes, carrying the matching genes for each pair, line up independently of each other. At the same time, two substructures in the cytoplasm, called *centrioles*, migrate toward opposite ends of the cell and begin generating fibers which fan out toward the chromosomes in the center of the cell. The fibers soon attach themselves to the centromere which joins the duplicated chromosomes. The fibers contract and pull the chromosomes apart and toward opposite ends of the cell until one of each duplicated chromosome has been collected around each centriole. A nuclear membrane begins to form around the chromosomes. The chromosomes lose their shape and resume their diffuse chromatin form. The nucleolus reappears.

Physical Division

The stage is now set for the creation of two identical daughter cells. A pinching action snips the cytoplasm in two and a new cell membrane forms to encircle the cytoplasm and its new nucleus. Mitosis is completed.

MEIOSIS

Meiosis is the cell division which guarantees the maintenance of the chromosome number when sexual reproduction takes place. Meiosis does this by producing gametes which carry only a single set of chromosomes—one of each chromosome pair. (Figure 19)

Like mitosis, the chromosomal material replicates and concentrates itself into doubled chromosomes joined at the centromere. Unlike mitosis, however, the chromosomes do not line up randomly in reduction division. Instead, as they contract, chromosomes carrying the matching genes seek each other out and wrap around each other. This is the process called synapsis we referred to in Chapter Four, and it is at this moment that crossover may occur. The chromosomes unwrap by the time they reach the center of the cell, but they line up next to each other, rather than randomly.

The fibers radiating from the centrioles pull the pairs of doubled chromosomes apart and toward opposite ends of the cell. The two daughter cells produced at this stage of meiosis have already undergone reduction division. They each carry only one doubled chromosome from each pair—or two copies of half the genetic material of the mother cell.

After a brief pause, these daughter cells each go through another division which is essentially mitotic in nature, producing a total of four daughter cells in all. Spindle fibers pull the doubled chromosomes apart and toward the centrioles. The cytoplasm separates, forming two cells containing a single copy of half the original mother cell's genetic material.

In sexually mature male animals, meiosis occurs in the testes. The four cells produced by meiosis are equal in size and next go through a process called *spermatogenesis.* (Figure 20) They lose most of their cytoplasm and develop the sperm's characteristic tail which provides the necessary propulsion to reach the egg cell.

In female animals, meiosis takes place in the ovaries, where the daughter cells go through the process of *oogenesis.* Unlike sperm, egg cells are not all the same size. The first meiotic division produces one large egg cell and a smaller cell called a *polar body.* The larger cell divides again and

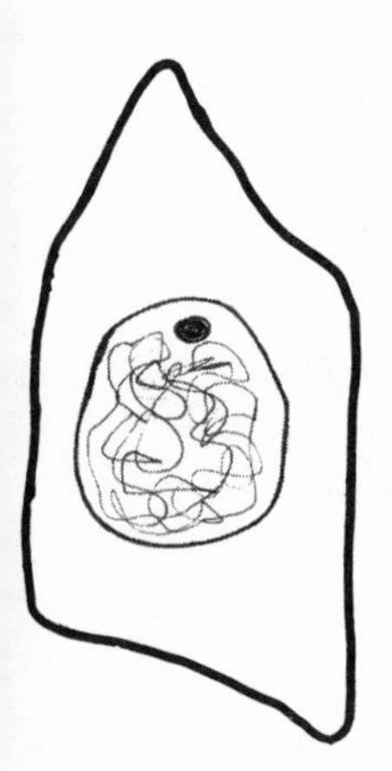

1. Chromosomes double in preparation for division

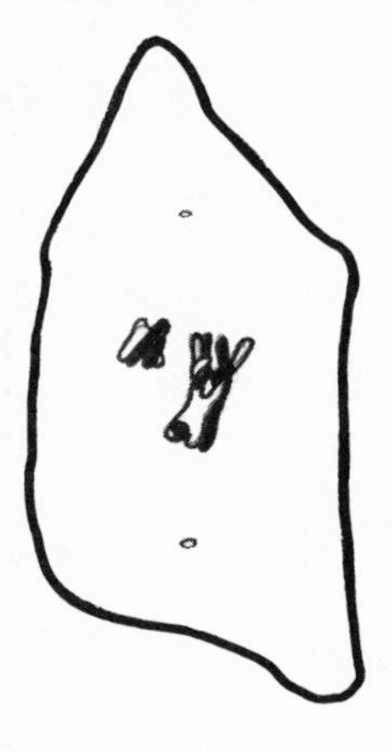

2. Doubled and paired chromosomes contract

3. Paired chromosomes line up on opposite sides of equator.

4. Chromosome <u>pairs</u> separate

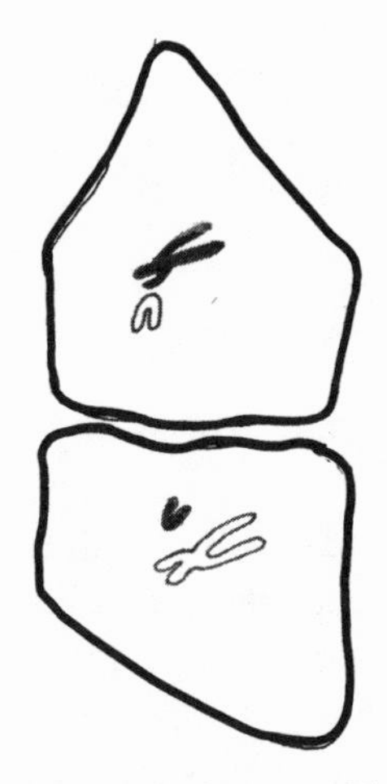

5. Daughter cells have half the original number of chromosomes

Fig. 19 Meiosis-Reduction division

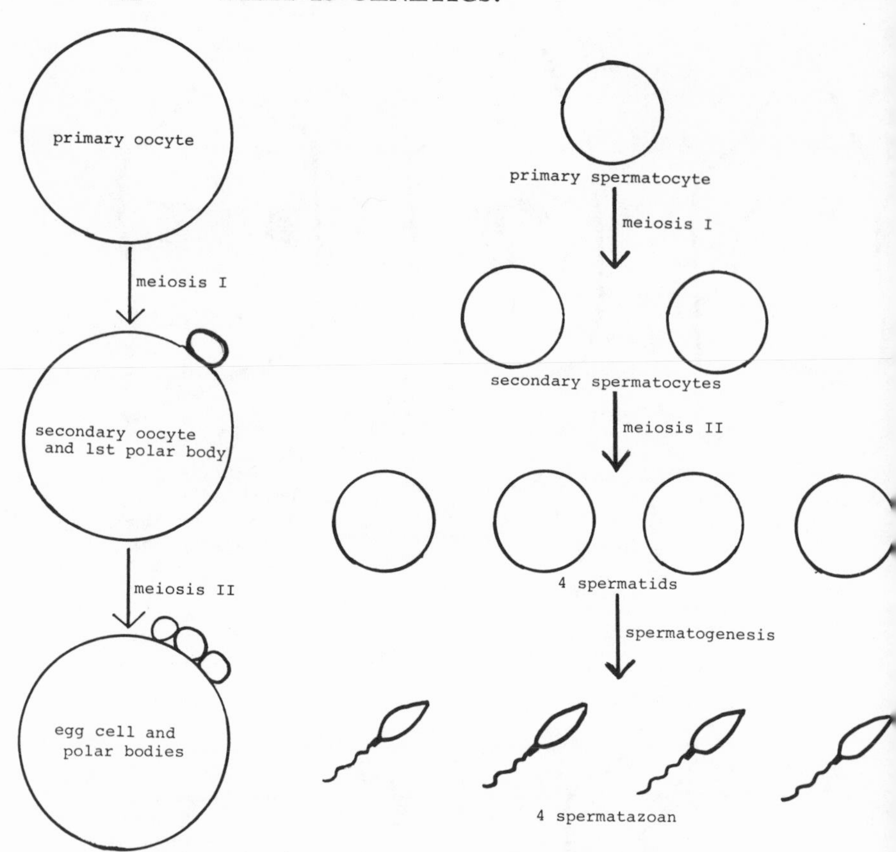

Fig. 20 Spermatogenesis and oogenesis

gives rise to a large egg cell and another polar body. The first polar body may or may not actually divide again, forming two more polar bodies. The final result is one large egg cell and three polar bodies which eventually disintegrate. Nature concentrates as much cytoplasm as possible into a single egg cell to guarantee an adequate food supply for the new living thing created through fertilization of the egg.

CHAPTER SIX

Is It a Boy or a Girl?

We live in a time when many people are actively redefining the old rigid sex roles, but sex is still the first thing we think about when a baby is born. The first question people ask about a newborn infant is not, Is it healthy? but, Is it a boy or a girl?

Strange as it may seem, the answer is not always easy. The sex organs of a newborn baby may be larger or smaller than usual. A boy's testes may not yet have dropped from the body cavity, making it difficult or unwise for a doctor to make a hasty announcement. The first judgment by the doctor in the delivery room is critical in determining a range of reactions that people will have toward the new baby—from the color of the blanket the infant will be wrapped in when it goes home from the hospital, to the type of clothing friends and relatives will buy as presents, to the name its parents choose to give it, to the way in which people will look at it. Sex roles may be changing, but the sex of the child is as important as it ever was.

Historically there has been a premium on having baby boys. This sexist bias has lost a lot of ground in the last decade, but there are many vestiges of pro-male prejudice in society today, including the passing of family names from father to son and the inheritance of family property. In the Middle Ages it was even worse. A king would often get quite angry with a queen who failed to bear him at least one male heir to his throne. Royal wives who had the misfortune of repeatedly giving birth to baby girls might be cast aside, or worse yet, even executed. As recently as 1948, when people should have known better, the King of Egypt and the Shah of Iran divorced their respective wives for failing to bear a male child to carry on the royal family line.

How sex is determined was unknown until quite recently, but as we have seen, it takes more than ignorance to keep people from conjuring up explanations. At one time or another it was believed that sex was determined by

——the phase of the moon at the time of conception

——the age of the parents at the time of conception

——whether the sperm fertilizing the egg came from the left or right testicle

——which ovary produced the egg—the right for boys, left for girls

——whether the baby developed on the right or left side of the uterus.

Incantations, rituals and special potions were concocted to influence the sex of the unborn child. In the Middle Ages, a pregnant queen might have found herself being told to lie in bed on her right side, holding her hands out with thumbs stuck up, if she wanted to have a baby boy. The king might have retained an alchemist to brew her a potion from the blood of a lion's heart, an eagle's head and the fleshy, wrinkled skin that hangs from the throat of a rooster, mixed with a specially blessed wine. Regardless of what they put Her Majesty through, no matter how many times they waved a magic wand and said abracadabra, the baby was just about as likely to be a girl as a boy.

HOW SEX IS DETERMINED

The moment the sperm fertilizes the egg, the new organism embarks on a path toward development *either* as a boy or a girl. Like all systems of development, this is a very complicated process with many phases and steps. The genes you receive from your parents will determine whether you develop ovaries or testes during the fetal period. These sex organs in turn determine whether male or female hormones will be produced to influence and control further sex differentiation during the gestation period. This will ultimately affect what you look like at birth and determine how you are identified and treated by those who know you, and how you are raised by your family. As a teenager, the very same ovaries or testes that were determined by your genes, will produce more hormones which influence your sexual maturation, body changes and sexual functioning. At the very end of this process is your own sexual identity—how you think and feel about yourself as a man or a woman. (Figure 21)

It is certainly possible for something to go wrong at any of these steps along the way that might alter the final result and make it something other than might normally be expected. But incantations and magic potions can never alter the fact that the baby was given its chromosomal sex determinants at the moment of conception.

The first hints about the real nature of sex determination came from: 1) the empirical observation that there is a general tendency toward a one-to-one ratio between the sexes in all species and 2) an understanding of Mendelian genetics. Scientists noticed that this one-to-one ratio was exactly what

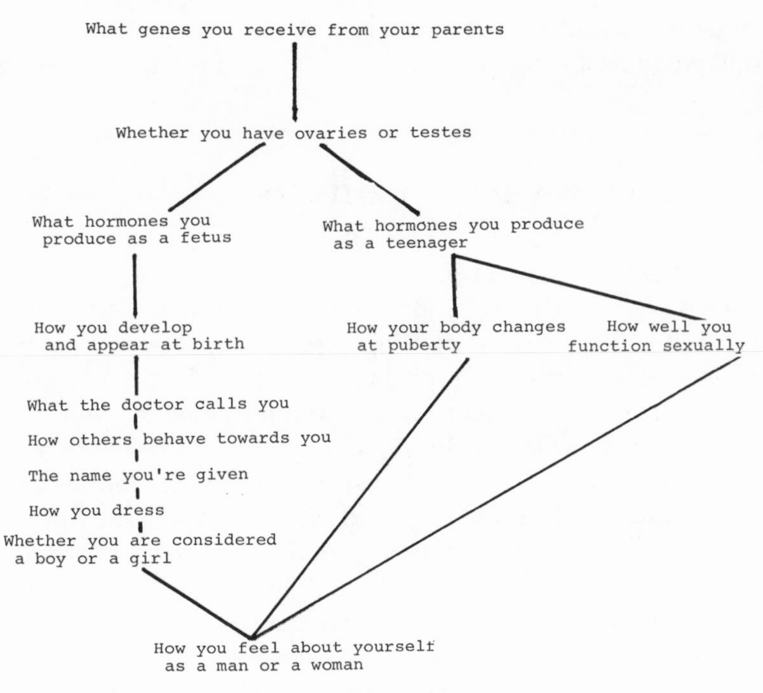

Fig. 21 How sexual identity develops

you'd expect when you crossed individuals pure for a recessive trait with individuals who were hybrid. (Figure 22) Microscopic investigation confirmed this suspicion. Twenty-two of the 23 pairs of human chromosomes were just as likely to be found in the cells of males as females. These were called the *autosomes*. The twenty-third pair, called the *sex chromosomes*, was different. In females, this pair was composed of two large chromosomes, which were named the **X** chromosomes. In males only one **X** chromosome was present. The other chromosome was a small truncated chromosome called the **Y** chromosome. (Figures 23 and 24)

These observations led to the chromosome theory of sex determination:

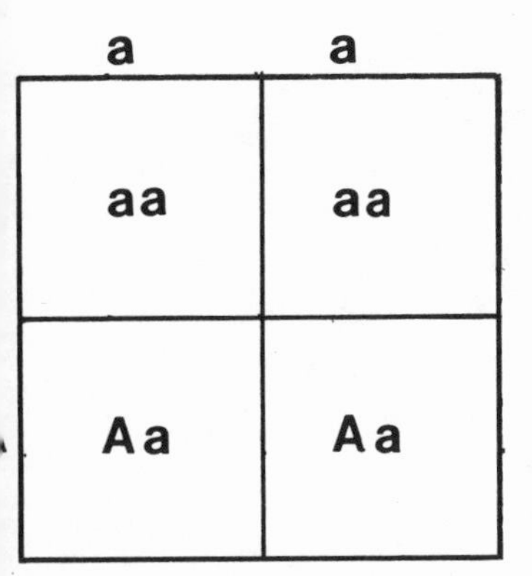

Distribution of phenotypes in crossing of pure recessive and hybrid. Ratio is 1:1, recessive to dominant phenotype.

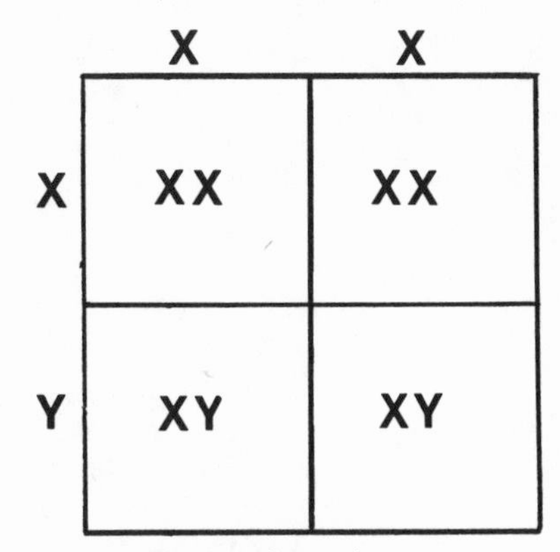

Distribution of sex chromosomes and resultant phenotypes. Male (XY) to female (XX) is 1:1.

Fig. 22 Sex determination

1) Sex is determined by the sex chromosomes.

2) The sex chromosomes in females are identical. Both are **X** chromosomes.

3) Since females have an **XX** genotype, all egg cells will carry an **X** chromosome.

4) Males do not have identical sex chromosomes. One is an **X** and one is a **Y**.

5) Since males have an **XY** genotype, half the sperm cells will carry an **X** chromosome and half a **Y** chromosome.

6)The fertilized egg cell may receive either an **X** or a **Y** chromosome from the sperm. Since all eggs carry an **X** chromosome, the chromosomal sex of the baby is totally dependent on the sperm. If it carries a **Y**, the baby will be a boy (**XY**); if it carries an **X**, the baby will be a girl (**XX**).

Those medieval prayers, magic spells and exotic potions were all aimed at the wrong patient! The queen kept having baby girls only because **X**-bearing sperm cells kept fertilizing her egg cells.

What we have said here applies to human beings pri-

CHROMOSOMES OF A DIVIDING FEMALE CELL

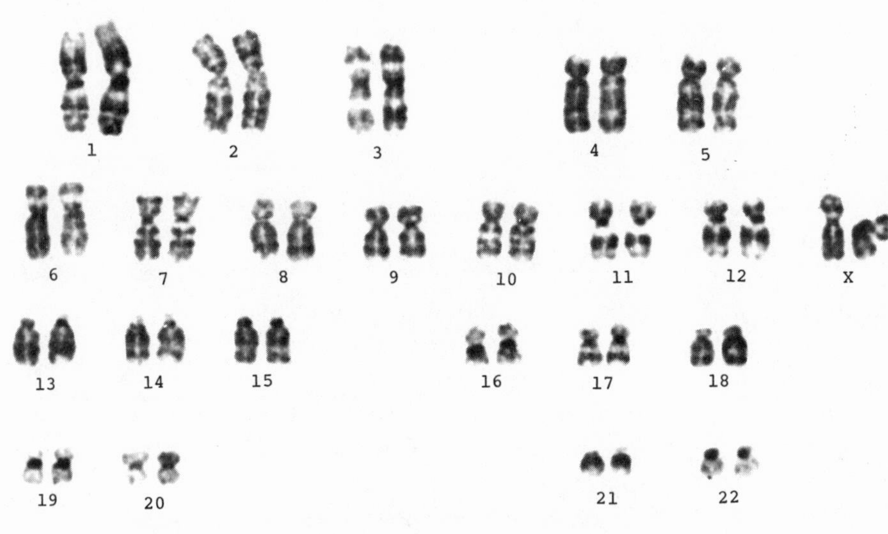

KARYOTYPE OF A NORMAL FEMALE

Fig. 23 Chromosomes of a normal female (Courtesy of the Cytogenetics Laboratory, NYU Medical Center)

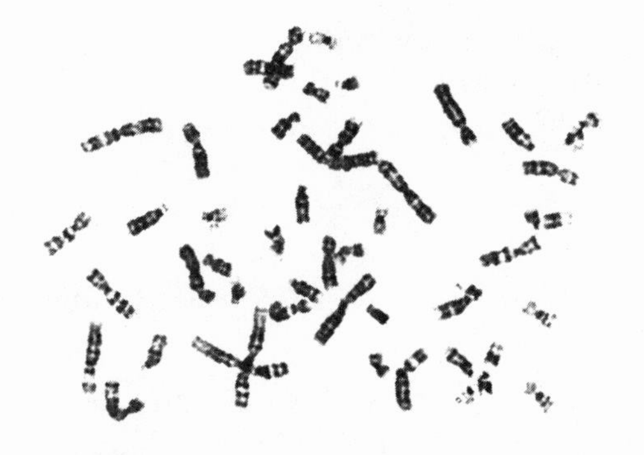

CHROMOSOMES OF A DIVIDING MALE CELL

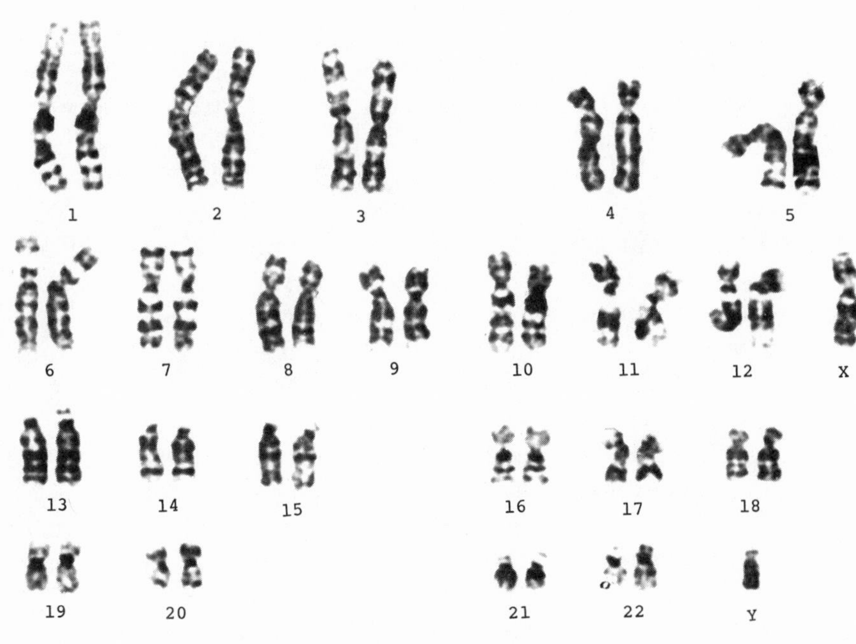

KARYOTYPE OF A NORMAL MALE

Fig. 24 Chromosomes of a normal male (Courtesy of the Cytogenetics Laboratory, NYU Medical Center)

marily. In some other animals and birds, identical sex chromosomes are actually characteristic of the male, and different sex chromosomes, of the female.

THE ROLE OF THE Y CHROMOSOME

The chromosome theory of sex determination was based on the simple observation that males and females had different chromosome configurations. The whys and wherefores of sex determination were still a mystery. Based on studies of laboratory animals, the genetic balance theory was widely accepted for many years as the solution to this mystery. The theory suggested that the large **X** chromosomes carried many genes for female traits and the tiny **Y** carried very few genes and was composed mostly of inert material. Genes for male traits were said to be dispersed on the twenty-two other chromosome pairs (the autosomes). If a cell had only one **X** chromosome, the male genes on the autosomes would outweigh the female genes on the **X** and tip the balance toward male development. If there were two **X** chromosomes, the female genes would outweigh the male genes on the autosomes and tip the balance in the other direction (Figure 25).

The first clue that this theory was wrong came from the study of human patients with sex chromosome abnormalities. Things go wrong sometimes and nature produces individuals with more or less than the normal number of sex chromosomes. There were patients who were found to have only a single sex chromosome—an **X**. This gave them an **XO** genotype. According to the genetic balance theory, the male genes on the autosomes should have overwhelmed the female genes on the lone **X** chromosome. The **XO** individual should have been a male. But **XO** individuals were always under-

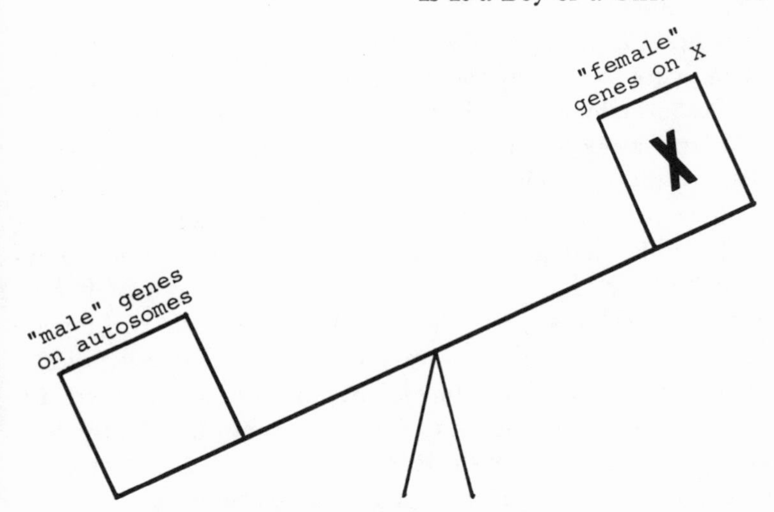

With one X present, "male" genes outweigh "female" genes. The result is a male.

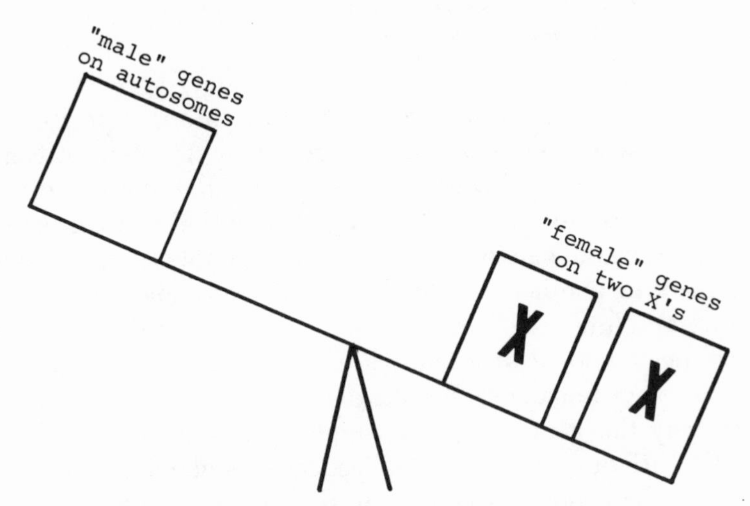

With two X's present, "female" genes outweigh "male" genes. The result is a female.

Fig. 25 The genetic balance theory of sex determination

developed females! Female genes on the single **X** chromosome were obviously powerful enough to counterbalance whatever male genes—if any—were found on the autosomes.

In still rarer cases, individuals were discovered who had three or more sex chromosomes—a **Y** and two, or even three, **X** chromosomes. Again, what was observed completely contradicted the balance theory. Individuals with double (genotype **XXY**) and triple (**XXXY**) the normal dose of female genes would have been expected to be pushed toward female development. But they were always males, though with a disrupted normal development. Far from being an inert, innocent bystander, the tiny **Y** chromosome apparently possessed the power to push development toward maleness, no matter how many **X** chromosomes were present.

SEX DETERMINATION VS. SEX DIFFERENTIATION

Sex *determination* is a once-in-a-lifetime proposition. It refers to the sex chromosomes you get at the moment of conception. Once fertilization takes place, the chromosomal makeup is set for life—**XX** or **XY**. Nothing that happens later can change the genetic composition of the unborn child from one sex to another. While the word determination sounds quite absolute, it is very important to keep in mind that future development of an individual will affect the degree of maleness or femaleness he/she displays. Sex *differentiation* is the way the genetically determined sex becomes a reality. Differentiation translates the coded message for sexuality within the genetic program into the physical traits, external and internal organs, bodily functions and behavior we normally associate with a person's sexual identity. Differentiation starts at conception—the moment sex is determined—

and continues through infancy, childhood, adolescence and adulthood.

Human beings have an inherent tendency toward female development. The absence of the **Y** chromosome results in female development. The mere presence of the **Y** unleashes a chain of events that switches development to the male pathway. It allows the formation of the testes, the male sex organs, to occur. The testes, in turn, produce two special hormones which trigger further male development. One of these hormones stimulates the formation of male ducts and external sex organs. The other hormone inhibits the development of female sex organs. Genes on the autosomes may or may not participate in sex differentiation, but the **Y** chromosome plays a fundamental role.

THE WEAKER SEX

We said earlier that there is a general tendency toward a one-to-one sex ratio in every species. This has to be clarified. Human males produce equal numbers of **X**- and **Y**-bearing sperm cells. This should logically result in a one-to-one sex ratio for each generation, since fertilization is random—but it doesn't. Birth statistics in the United States and Europe, where reliable statistics have been kept for a long time, show that about 105 boys are born for every 100 girls year after year. For a long time this discrepancy was used to prove the old cliché about girls being the weaker sex. Boys were inherently stronger and therefore more likely to survive the hazards of pregnancy and actually be born.

This is another folk legend that just isn't true. Scientific evidence suggests just the opposite. Boys are less likely to survive pregnancy than girls. Male fetuses who die and abort spontaneously during the first few months of pregnancy outnumber girls by almost three to one. In fact, males have a higher death rate than females at every age level, from the moment of conception through pregnancy, childhood and adulthood. Life expectancy for males is less than females. If there is a weaker sex, it is the male of the species. (Figure 26)

If this is true, why do more boys enter the world kicking

Fig. 26 Table of death rates by sex

AGE	MALE	FEMALE
Under 1	1,830	1,444
1 - 4	78	64
5 - 9	42	29
10 - 14	46	26
15- 19	147	54
20- 24	210	67
25 - 29	200	75
30 - 34	206	98
35 - 39	277	146
40 - 44	419	237
45 - 49	667	366
50 - 54	1,044	544
55 - 59	1,615	821
60 - 64	2,523	1,227
65 - 69	3,636	1,731
70 - 74	5,556	2,945
75 - 79	8,254	4,879
80 - 84	11,593	7,687
85 and over	17,573	14,031

(per 100,000 population)

The prenatal death rate is three times greater fo boys than it is for girls. This higher death rat for males continues throughout life at every age level. (Based on statistics from the United Stat Center for Health Statistics.)

and screaming than girls? The explanation appears to be that more boys are conceived, perhaps as many as 130 boys for every 100 girls. Because the male prenatal death rate is so high, the ratio drops to 105-to-100 by the end of the prenatal period. The adult male produces equal numbers of **X**- and **Y**-bearing sperm, but there is something about **Y**-sperms that makes them more likely to fertilize the egg. The difference in size and mobility seems to be the most plausible explanation. Since the **Y** chromosome is so much smaller than the **X**, the **Y**-bearing sperm can swim faster and more successfully in the journey to reach the egg cell.

Prenatal environmental conditions also influence the chances of any fetus—male or female—to survive pregnancy. Favorable conditions encourage normal development and the birth of a healthy baby. Unfavorable conditions are apparently more likely to hurt the survival chances of a male fetus than a female. The mother's age and health seem to be crucial elements of the prenatal environment. A young, healthy mother carrying her first baby is more likely to have a boy than an older woman in poor health. For younger mothers, the ratio of boy-to-girl babies may get as high as 120 to 100. Among mothers over 40 years old, the ratio may drop as low as 90 to 100. The key thing here is not simply maternal age, but general overall health which is linked to the aging process.

A long-standing popular folk tale used to tell us that nature acted to increase the male birth rate after a war, so as to compensate for the deaths of so many male soldiers. Birth statistics certainly show an increase in male births after such catastrophes as World Wars I and II, but the explanation is not some mysterious natural intervention to restore population balance. Natural processes hardly operate in a conscious manner. During war, young men are away from home in the military in large numbers. Marriages and families are postponed. When the war ends, the men return home. There is a sudden jump in the number of marriages. A baby boom follows shortly after. Many of these new mothers are young women, the most likely to bear boys. If there hadn't been a war with its disruption of normal marriage and birth trends, these women would have had their first children spread out over a number of years. But because of the war, they have their first babies—the ones most likely to be boys—concentrated in a postwar baby-boom period.

Sex has been and always will be the most important trait you can inherit. But the sex chromosomes are involved in more than just sex determination. They carry genes that relate to other traits as well. The fact that boys and girls don't have identical sex chromosome complements has some very significant implications for the inheritance of these nonsex characteristics.

CHAPTER SEVEN

Sex-linked Traits

Queen Victoria became the queen of England in 1837 and ruled for sixty-four years. She and her husband had nine children, four sons and five daughters. One of the boys, Leopold, suffered from a tragic disease called hemophilia. A hemophiliac's blood lacks a key ingredient required for the blood to clot. Until quite recently the slightest scratch or cut put the hemophiliac in danger of bleeding to death. Few victims survived childhood. Even today, despite medical breakthroughs which prolong life expectancy for victims of the disease, there is no cure for hemophilia.

Hemophilia is not contagious. You cannot catch it; it is not caused by a germ. It is entirely hereditary, controlled by a recessive gene on the **X** chromosome. A person is born with it or without it.

Everyone probably carries at least three or four "defective" genes tucked away somewhere in one or another of their chromosomes. Defective genes which might cause a harmful form of a characteristic—like the gene for hemophilia—are usually recessive. The only way for them to express themselves in the phenotype is through combination with a similar recessive gene on the corresponding chromosome. This means both parents would have to contribute the same defective gene. Even then nature has provided certain safeguards which minimize the effects of defective genes. Since most traits are polygenic (controlled by several genes), the full negative impact of a defective gene is frequently blunted. Some polygenic traits are thought to be controlled by many genes, each with an equal, small effect on the overall expression of the trait. If one of these gene pairs is defective, it won't do much damage. Other traits may be controlled by one major gene pair and modified by the small effects of many others. How drastic an impact a defective pair might have would depend on whether it was the major one, or one of the minor ones.

SEX-LINKED GENES

The inheritance of hemophilia is not subject to the same safeguards. The concept of linkage groups applies to sex chromosomes, just as it does to other chromosomes. There are perhaps as many as a hundred genes located on the **X** chromosome which are inherited as a group. These are called sex-linked traits. Since they are located on the **X**, it would be more accurate to call them **X**-linked. Some very important traits are apparently controlled by a single gene pair found on the **X** chromosome, including the ability to distinguish colors and the ability of the blood to clot properly. In Chapter Six we pointed out that sex chromosomes differ from the autosomes in that they are not identical and interchangeable. The **X** and the **Y** do not carry matching genes. This situation holds important consequences for the transmission of sex-linked genetic diseases like hemophilia.

Males have an **XY** genotype. Non-sex-related genes appearing on the **X** chromosome will express themselves in the phenotype—even if it's a recessive gene—because the **Y** doesn't have any gene capable of masking it. Prince Leopold was born with hemophilia because his **X** chromosome carried the defective gene that causes the disease. There is only one place he could have gotten his **X** chromosome from—his mother, the queen. (Remember that all egg cells have an **X** chromosome.) If he lives long enough to reproduce, a hemophilic male cannot pass the disease to his sons, though his grandsons may very well be victims. His daughters will not normally suffer from the disease, but they will be carriers who can pass it on to their sons. Because he was born into the royal family, Leopold received sufficient care to allow him to survive to adulthood, to marry and to have a child. Leopold had a daughter who was a carrier. Her son was born a hemophiliac. Why and how this every-other-generation pattern occurs can be understood by means of Mendelian genetics.

We can designate an **X** chromosome with the defective hemophilia gene as **Xh** and an **X** chromosome with the normal gene as **Xn.** A male hemophiliac like Leopold would have to have an **XhY** genotype. A normal woman like Leopold's wife, would have an **XnXn** genotype. The hemophiliac father would pass on a **Y** chromosome to every son and the normal mother an **Xn**. Thus, every son of this couple would have an **XnY** genotype and would not suffer from the disorder. (Figure 27)

Each daughter would receive an **Xn** chromosome from the normal mother and an **Xh** from the father. Thus every daughter would carry an **XnXh** genotype. Since the defective gene is recessive, the normal gene on the **Xn** would dominate and the female would not suffer from the disease. Since recessive genes are only masked—not changed or destroyed in any way—they may reappear in the phenotype of future generations. The **XnXh** female is a carrier of the disease.

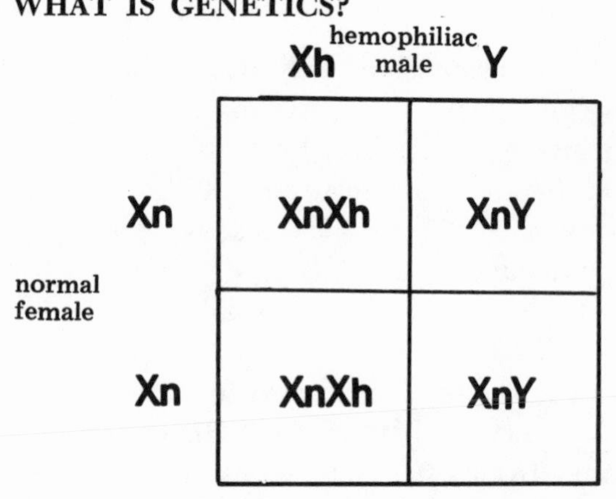

Fig. 27 Theoretically possible offspring from union of hemophiliac male and normal female.

Even if the **XnXh** woman marries a normal male, fifty percent of her sons are at risk of being hemophiliacs. Half her egg cells will carry the defective **Xh** chromosome and half the **Xn**. The normal father will give every son a **Y** chromosome. Fifty percent of the boys will be **XhY**. Since there is no gene on the **Y** capable of masking the defective gene, the **XhY** sons will suffer from the disease. (Figure 28)

For many years it was believed that hemophilia never occurs in females. You may have heard or read that sex-linked diseases occur only in males. This is another one of those half-truths which obscure how heredity works. The source of the confusion is the fact that males who suffer from hemophilia don't usually survive long enough to reproduce. For a girl to suffer from the illness, she'd have to have a recessive-defective gene on both her **X** chromosomes. Her mother would have to be a carrier and her father, a hemophiliac. Since most hemophiliacs have historically died young, the matings that could produce a female hemophiliac have been very rare. Several such matings had been observed, but the offspring never included an afflicted female. This led to specu-

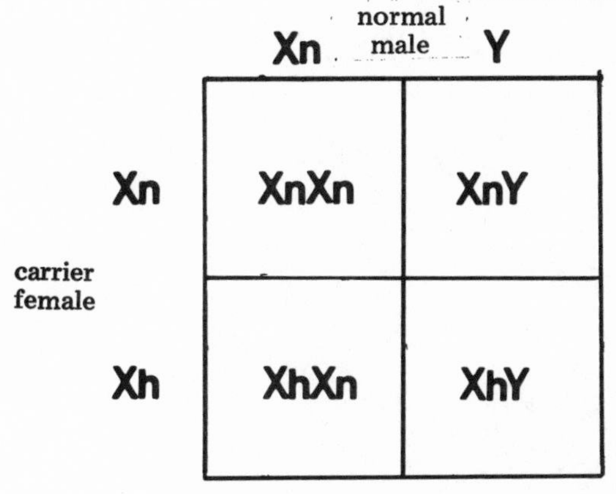

1/4 normal females: 1/4 carrier females: 1/4 normal males: 1/4 affected males.

Fig. 28 Theoretically possible offspring from union of normal male and "carrier" female.

lation that a female hemophiliac was impossible for a variety of theoretical reasons—all of them erroneous. In the 1950s, the first two proven cases of hemophilic women were discovered, and another false theory was discarded. Female hemophiliacs are rare simply because the mating that can produce them is rare.

When the defective sex-linked trait doesn't have lethal effects, and the males live to reproductive age, the trait appears more frequently in women (although still much less frequently than in males). Color blindness is a notable example. The central criterion for identifying a sex- or **X**-linked trait is the inability of the affected man to transmit it directly to a son or to a future generation through a son. It is not the mistaken view that sex-linked diseases strike only males.

When the sex-linked genetic disease is so destructive that it causes males to die before reproductive age, the overwhelming likelihood is for the defective gene to die out quickly within a few generations, as the family tree of Queen Victoria illustrates. (Figure 29) It may be passed on by carrier sisters for several generations. But it has been estimated

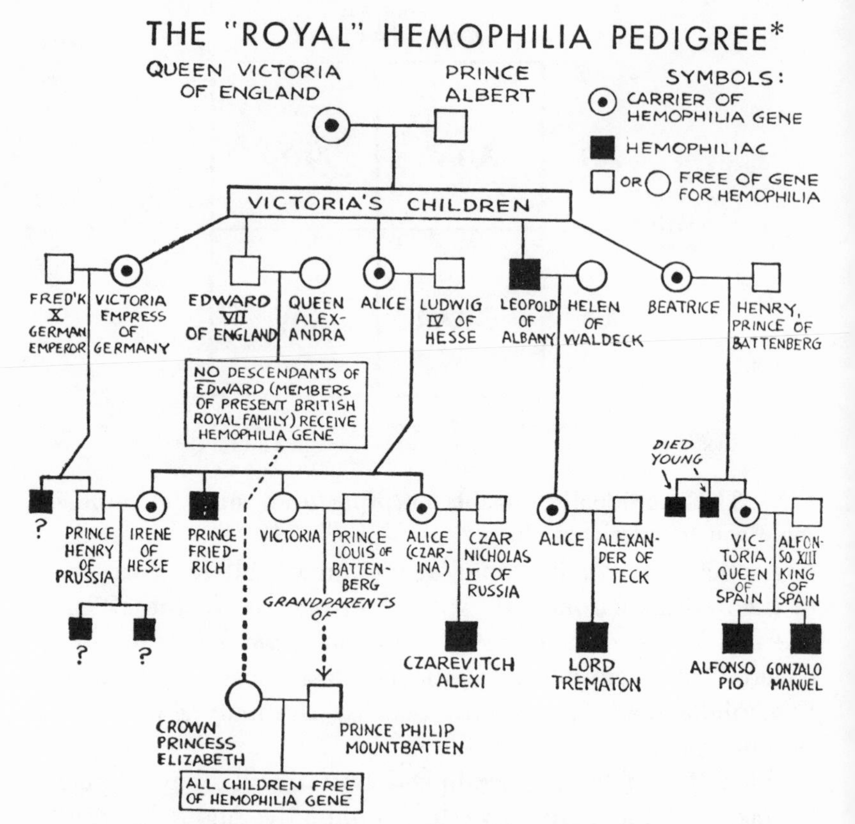

*Only some of the members of each of the families represented are shown here. The Chart is based on a diagram and data compiled by Dr. Hugo Iltis. (*Journal of Heredity*, April, 1948.)

Fig. 29 The Family Tree of Queen Victoria (*From Your Heredity and Environment*, by Amram Scheinfeld, copyright © 1965, 1950, 1939 by Amram Scheinfeld. Reprinted by permission of J. B. Lippincott Co.)

that the life span of such lethal genes is often as little as three generations.

If lethal genes tend to die out so quickly, why don't they disappear altogether? Why do hemophilia and other sex-linked diseases, like Duchenne's muscular dystrophy, persist in the human population? The answer is: mutation. Mutations are chemical changes in genes that may occur more or less randomly in nature all the time. Most hemophilia patients are usually the first known victims in their families. Leopold was the first victim of hemophilia in the royal family. The fact that her daughters were carriers who subsequently gave birth to hemophilic sons indicates that a mutation must have occurred within Queen Victoria or one of her parents altering the gene for clotting to its defective form.

No discussion of sex-linked conditions would be complete without mention of the **Y** chromosome. The **Y** is tiny and has room for very few genes. For years it was thought that there were a number of **Y**-linked traits, other than maleness. The most famous was the "porcupine skin" trait which was cited in many genetics textbooks. Recent research has, however, proven there is only one **Y**-linked trait we can be sure of, aside from the one triggering the development of maleness. This is the hairy ears trait found in parts of the Indian subcontinent. This trait is picturesque but in no way detrimental to health. Since the gene is found on the **Y** chromosome, the trait can only appear in males.

CHAPTER EIGHT

Evolution and Genetics

Until a little more than a hundred years ago, the biblical version of creation was accepted as Revealed Truth by all but the most skeptical free thinkers. Thanks to the patient work of James Ussher, a seventeenth-century Irish churchman, millions of people believed that the planet Earth was 6,000 years old. The archbishop had taken it upon himself to cull through the Bible, keeping track of who begat whom and adding up the ages of everybody mentioned. In 1654 he announced the conclusion of this study: God had created the Earth at 9 A.M., October 26, in the year 4004 B.C.

Today we know enough about the history of the Earth to realize that six thousand years is closer to yesterday morning than it is to the beginnings of the world.

Geology is the science that studies the structure of the Earth's crust and its layers, and it was in this field that the first breakthroughs were made in appreciating the antiquity of the Earth in the early 1800s. Certain pioneers in geology pointed out that the earth is not a static place. River water surges through rocky terrain, eventually carving out canyons. Waves pound against shorelines and erode beaches. Currents and tides carry and deposit sand, leaves and other organic material in marshes. Gradually these decay and new land may be built up where once there was none. The Earth is in a constant state of change. Changes usually take place at an extremely slow, almost imperceptible pace. But there is constant change and it can be measured.

The exact amount of sediment deposited by a river in a single year could be measured directly. If sediment accumulated at a uniform rate, it was simple to calculate how long the process had been going on. Estimates of the Earth's age suddenly jumped from 6,000 to 100 million years.

To the creationists, who believed in the Bible's version of creation, the idea that the Earth was 100 million years old was heresy. They believed that the world was created in six days. All the physical features of the planet and all the forms of life that inhabit the Earth had not changed since creation, in their view. Fortunately, their fervent denunciations didn't block further scientific inquiry. We realize now that even the 100-million-year figure is a gross underestimate. The generally accepted estimate today is in the neighborhood of 4.6 billion years. Life, however, is not nearly as old. Earth was probably formed by the gradual condensation of clouds of hot gases. The temperature of these gases was probably in the thousands of degrees. Living things would have sizzled and fried under such conditions.

It probably took at least a billion years just for the gases to cool and solidify enough to form the Earth's crust. And even then life was aeons away. The very beginnings of life, the accidental formation of amino acids, which are the building blocks of proteins, probably began about 2.5 billion years ago. The first cells didn't form until 1.5 billion years ago.

A British naturalist named Charles Darwin watched the debate about the Earth's age with great interest. Darwin had been working on a theory of evolution that would explain how species developed and changed over time. His theory could even explain how new species originated. There was only one problem. The process Darwin visualized couldn't possibly have had enough time to function if the world were only six thousand years old. When geologists suggested that the world was perhaps 100 million or more years old, Darwin was encouraged.

Darwin began work on his theory to make sense out of what he saw during an around-the-world voyage as ship's naturalist for the *HMS Beagle* in 1832. He had been particularly intrigued by the strange animals and birds he had seen during a stopover at the Galápagos Islands, located near the equator, about six hundred miles west of Ecuador, in the Pacific Ocean. The strange species he saw included penguins (which most people even today mistakenly believe are found only in the Antarctic), a rare type of the seabird, the cormorant, which couldn't fly, and a rare breed of mockingbirds. He also discovered twelve different types of finches. The types varied from island to island in the Galápagos chain. He believed that these finches, which have since become known as the Darwin finches, must have descended from a common ancestor. He suspected that after they had become geographically isolated from each other, and after years and years of separate development, the finches had gradually evolved into separate species. They were incapable of interbreeding. The question in Darwin's mind was how and why had this happened?

Darwin had to begin with the accumulated knowledge on hand. The central elements that laid the basis for his theory were: his knowledge of animal breeding, the research of the French biologist Jean Baptiste Lamarck on the ability of plants and animals to adapt to their environment, Darwin's own observations of the existence of variation in and between species, and an essay by Thomas Robert Malthus.

From his knowledge of animal breeding, Darwin knew

that people have been able to change and modify species of domestic animals for hundreds of years. Dairy farmers had developed breeds of cattle that were good milk producers by carefully selecting and mating their best milkers. Breeds of superior egg-laying hens had been developed in a similar manner. Racing horses had long been bred for speed. It was clearly possible for man to change a species by *artificial selection* of those individuals who would be allowed to reproduce. In so doing, farmers controlled what characteristics (milk producing, egg laying, speed, etc.) would be passed on to future generations. Perhaps, Darwin thought, nature had her own way of selecting the characteristics that would be inherited.

From T. R. Malthus' essay on population, Darwin got an idea about how *natural selection* might work. Malthus said that history showed that the human population increases faster than the size of the food supply required to feed it. Overpopulation had historically been prevented only by periodic wars, epidemics and famine. Many people die without reproducing—without leaving any children. It sounds quite grim, and many writers have quarreled with the social implications of Malthusian theory which leave the door open for regarding such social calamities as war and famine as blessings in disguise. But Darwin saw certain insights in Malthus that he could apply to his own efforts at hammering out a theory of evolution.

Darwin noted that individual organisms in every species have the potential to produce many offspring. Female fish lay thousands of eggs at a time. A female mouse can have several litters of 12 to 15 babies per year. Each offspring possesses a similar reproductive potential. A single pair of mice producing three litters of 15 babies can give rise to 675 descendants by the grandchild generation, or 2,000 within two years.

Despite this fantastic potential for population growth, and

overpopulation, the natural population of any species remains fairly constant over time. From time to time there may be variations in population levels because of natural calamities and upheavals, or human interference in the environment. Floods, earthquakes or the coming of an Ice Age are illustrations of natural phenomena that might affect population levels. The slaughter and near-extinction of the American buffalo or the devastating effects of industrial pollutants on plant and fish life are examples of how human interference can upset a species' normal population level.

But left undisturbed, nature usually maintains the population of a species at a constant level. This happens because many members of each generation do not reproduce, or reproduce significantly below their potential. Some die young, before reaching reproductive age, or before completing their full reproductive period. Some are not healthy enough to reproduce.

Darwin was also aware of the tremendous variation within every species. He believed there was a very important connection between these variations and the likelihood that a particular individual organism would reproduce. Just as the farmer chooses only his best milkers to reproduce, so nature must have had a way of choosing which variations were passed on to future generations.

NATURAL SELECTION

Darwin concluded that the natural environment that organisms live in acts as a natural selector. Over long historical periods those variations get passed on which help individuals to be better suited to survive in the environment. Animals in a particular place may confront extremes of heat and cold, limited food supply and predators as threats to survival. Those individuals possessing variations which help them to surmount these challenges live longer, stay healthier and are more likely to reproduce. Over time an increasing number of new generations will have these adaptive traits. Over time the inferior traits—those inadequate for survival—will tend to disappear from the gene pool available for the species.

In the severe cold of the Arctic region, a large-bodied Siberian husky, born with a thick coat, is well-adapted to survive in that inhospitable environment. It has a good chance to survive and participate in producing the next generation. A small, hairless or short-haired dog would likely freeze to death in that climate. It would be ill-adapted for survival. Its coat would not provide protection against the cold, and small bodies tend to lose body heat at a faster rate than larger bodies.

Another Arctic animal, the polar bear, possesses two adaptations that improve its chances to survive in that environment. Its thick white fur provides warmth and camouflage in snow and ice which improves its ability to hunt and feed itself. At the same time, under the white fur, it has a black skin capable of doing a good job of absorbing heat from the sun. The polar bear's white fur, which might ordinarily reflect sunlight, will actually stand up on end and permit the black skin to be exposed to the sun's heat!

In the warmer climate of Africa, giraffes offer another frequently cited example of successful adaptation to the environment. Giraffes were not always the long-necked creatures they are today. At a certain point in history, because of environment conditions, giraffes began eating leaves on trees as a source of nourishment. As competition with other species for this food supply on the lower leaves increased, those individual giraffes who had longer necks were better equipped to reach leaves higher up in the trees. These longer-necked giraffes ate better because they didn't have to compete with other species for food. They were better adapted to survive and therefore to reproduce. Short-necked giraffes were less able to find food, less healthy and faced possible starvation. The genetic material which gave rise to the potential to develop long necks was more likely to be contributed to the next generation than the genetic material for short necks. This process of natural selection operated over millions of years to produce a population of giraffes with long necks.

Successful adaptations or variations are perpetuated in a species and tend to become incorporated into the normal or modal phenotype for most individuals in the species. Variation persists but it ranges around an evolutionarily determined norm. Today giraffes' necks still vary in length, but more toward the long end of the spectrum. Hair thickness varies from husky to husky, but toward the thick end of the possible spectrum.

It would be a mistake to think of natural selection as a Grim Reaper, ruthlessly deciding who lives and who dies, which genes are passed on and which are discarded. Natural selection differs from artificial selection in that it is not a conscious process. The raw material natural selection has to work with is determined largely by chance. Evolution consequently occurs over extremely long periods of time. This is just as well because sudden, dramatic shifts in the overall phenotype are incompatible with the finely balanced nature of the genetic program. Farmers can achieve changes in key traits in their livestock fairly quickly by using consciously directed artificial selection. But this usually happens at the expense of proper genetic balance and overall well-being of the breed. Domestic breeds are frequently incapable of independent existence away from the controlled environment of the farm.

It would also be a mistake to think that natural selection leads to uniformity (lack of variation) or perfection. The greater the variability that persists in a species, the better are the chances that it will continue to adapt to changes in the environment in the future and to thrive. Those species that become highly specialized become very restricted in their ability to respond to environmental change and face extinction, like the dinosaurs. Because the world is always changing, it is impossible to say that evolution stops or that perfection is achieved. The evolution of insects and rodents which are resistant to pesticides that human beings have developed to kill them is an example of how this process continues.

It is important to keep in mind that natural selection acts on phenotypes, not genotypes. It is not a process that throws out individual genes and keeps certain other ones that might produce traits better suited for survival. As we saw in our discussion of Mendelian genetics, the same phenotype may result from different genotypes, even if the trait is controlled by a single pair of genes. (The pure, tall plants and the hybrids both exhibited the same phenotype.) For polygenic traits, numerous different combinations of gene pairs may produce the same or closely similar phenotypes. Several different phenotypes may be beneficial or even neutral in terms of survivability in a given environment and there is no reason for such variations to be eliminated by natural selection.

Natural selection doesn't eliminate genes that cause ill-adapted traits quickly or totally. Though some maladapted individuals may die young, or reproduce less, some of them will reproduce and contribute their genes to the next generation. Whatever weeding out takes place occurs over many generations. In addition, mutations occur repeatedly, temporarily reintroducing the same genes. Some seemingly harmful genes can also act to improve chances of survival under certain conditions and will be retained within the gene pool. For example, the gene that causes sickle cell anemia in people of African descent is a recessive gene. A person who is a pure recessive for this trait may die. But because of the particular way this gene operates, a person who is hybrid for this trait seems to have a selective advantage for survival in parts of the world where the disease malaria is a problem.

By the time Darwin had pieced his theory of evolution together, he felt he could explain the existence of the Galápagos finches. He concluded that the birds must have become geographically isolated from each other at some distant time in history. They pursued divergent paths as they continuously adapted to their particular environments. Beak sizes and shapes most suited for survival on the particular island gradually predominated and became characteristic for the population on the island. After sufficient evolution the finches became so distinct that they in fact constituted new and different species.

Geologists had demonstrated that the Earth had not been created as it is now. Darwin had theorized that life too was not created in one day in a permanently fixed way. Evolution continues all the time. New species come into being. As further evidence of his theory Darwin cited the surprising similarity in the bone structures of such different creatures as birds, animals and humans. The tiny mouse and the giant giraffe look as different as you can imagine, but they both have seven bones in their necks. Darwin thought this meant that all species must have evolved from a single evolutionary ancestor millions and millions of years in the past.

If the creationists had been in an uproar over geological theories about the Earth's age, you can imagine their response to the theory that suggested that God had not created all forms of life in one fell swoop. Moral outrage against Darwinian theory persisted into the twentieth century. The film *Inherit the Wind* which depicts the "Scopes monkey trial" in Tennessee, in which a high school teacher was tried for violating a law against teaching evolution, indicates the nature of the religious opposition to evolutionary theory. Even to this day many fundamentalist religious groups object to their children being taught about evolution. By and large, however, society today is more tolerant. Whatever moral and ethical value people may get from the Bible, it is no longer considered a source of scientific fact.

Creationists were not the only opponents of Darwinism. The scientific community quickly took sides for or against the theory. Even those who supported the concept of evolution disputed specific parts of Darwin's theory. Most notable were the followers of Lamarck (1744–1829), much of whose work on adaptation had served as a basis for Darwin's work. The Lamarckians had a different explanation for changes in species. They believed that *acquired characteristics* could be inherited. According to this view, early giraffes acquired long necks by constantly stretching to reach leaves higher up in the trees. These longer necks were passed on to their offspring.

The classical laboratory experiment to refute this Lamarckian theory was to lop off the tail of white mice and permit them to reproduce. This was repeated generation after generation, but new generations were always born with tails. And their tails were always as long as the tails of mice whose ancestors had never had their tails cut off. There are countless other examples that disprove this Lamarckian theory. A man or woman may have their tonsils or appendix removed, but their children are always born with these organs. Jewish boys have had circumcisions performed on them for thousands of years without any physiological changes—newborn Jewish boys are always born with foreskin. A weightlifter may acquire bulging biceps through constant exercise, but his children are never born musclebound. Except for environmental influences, like his father's emphasis on a good physique, there is nothing to rule out a weightlifter's son being a "97-pound weakling."

A far more serious criticism of Darwin charged that he couldn't explain the source and persistence of variation in a species. As mentioned in Chapter Two, the prevalent theory in the 1860s saw heredity as a blending process, in which parental traits were mixed together and appeared in a watered-down version in the offspring. The cornerstone of Darwin's theory was the incorporation and spreading of new variation throughout a species. But this was impossible if the blending theory were correct. New traits would be diluted by half in the first generation, by one-fourth in the next, by one-eighth in the generation after that, and so on. Rather than becoming increasingly widespread, new traits would be quickly diluted and either disappear entirely or become insignificantly rare in the population.

The answers to these criticisms were to be found in genetics, a science which had not yet come into existence in the 1860s. Unable to come up with answers, Darwin eventually retreated to a "sub-theory" he called *pangenesis*. This subtheory combined elements of ancient Greek reproductive theories and Larmarckian ideas about the transmission of acquired characteristics. Darwin suggested that each part and organ of the body prepares a small-scale model of itself which is transported by the bloodstream to the sex organs where they are combined to form gametes. In this manner, acquired characteristics, or new variations, also contributed small-scale versions of themselves which could be passed to new generations.

The answers to these criticisms could not be found partly because Mendel's work had been universally ignored. Mendel's experiments had disproved the blending theory and could explain how variation persists. Even those genes which do not express themselves in a particular generation don't disappear forever or get watered-down. They exist unchanged within an organism which may display a completely contrasting phenotype, and they can reappear unchanged in a future generation. Mendel's laws of Segregation and Independent Assortment showed that the tendency is to maintain the widest possible variation within a species and Darwin's work showed that the only limit to this variation was the ability to survive in the environment. Because Mendel's work had been ignored, Darwin couldn't draw on the valuable insights which buttressed his theory and so he remained completely stymied in his ability to respond to his critics.

As for the origin of variations, it wasn't until the twentieth century that geneticists would discover the significance of mutations. Today we realize that mutations are historically responsible for all genetic variation since the beginnings of life. These genetic changes are responsible for changes in the phenotype. Natural selection acts upon the raw material provided by these mutations to improve the ability of species to cope with a changing environment.

It is indeed a shame that Darwin could not draw on genetics to bolster his theory. He was so shaken by the vehemence of the criticism, and by the atheistic conclusions drawn by some of his followers, that some historians believe he died no longer believing in his own theory of evolution.

CHAPTER NINE

Mutations

Variation is the essence of life. The existence of variation within and between species was the underlying framework for the work of both Darwin and Mendel. Darwin constructed his theory of evolution based on the variation he could see everywhere in the world around him. Mendel explored the ways in which variation endures from generation to generation.

But neither of these great scientists could explain the source of variation. Darwin couldn't come up with an explanation. Mendel never even tried to deal with the question. He focused on characteristics which took obvious and distinctive forms and studied how they were distributed in the offspring of each generation. In so doing he arrived at certain basic laws governing how traits are shuffled and reshuffled in the hereditary process. But where did these different traits come from? If all pea plants had the same genes for height, there wouldn't have been anything for Mendel to study. If all pea plants carried genes only for tallness, all plants would have been six feet tall and that's all there would have been to it. Mendel figured out a pattern of inheritance for tall genes and short genes. But why were there tall genes and short genes in the first place?

Genes are tiny organic molecules. And though they're too small to be visible under even the most powerful microscopes, geneticists have been able to learn much about how they're constructed and how they function. Like all molecules, genes are subject to the laws of nature, and one of the most fundamental natural laws is that everything can change. If wind and water can whittle solid rock into a Grand Canyon, it shouldn't be too hard to understand that genes, which are living and dynamic, can change. These genetic changes are called *mutations*.

The importance of mutations cannot be minimized. Mutations are the source of all genetic variation. The genetic material present in the very first living cells had a truly remarkable ability to change. An incredible number and variety of alterations, rearrangements and duplications of genetic material have occurred in the past billion and a half years. These have yielded a wide diversity of forms, including the one-cell amoeba, tiny insects, giant dinosaurs, towering sequoia trees, and intelligent human beings. These divergent life forms did not result from anything being added to the genetic material from the outside. They stem from changes in the genetic material that was present in and reproduced from the very first living cells. Mutations have created all the different forms genes take within and between all forms of life. They have furnished the raw material upon which Mendelian laws, natural selection and evolution operate. Variation is the essence of life; mutations are the source of all variation.

CHANGES IN THE GENETIC CODE

Mutations are alterations in genetic material. They don't automatically or directly act on an individual's phenotype. The genetic material is the blueprint for the individual organism being constructed. It affects the individual's phenotype through the building blocks of life—the proteins. The proteins are synthesized within the cell according to the instructions coded in the genetic material. If the code is changed by mutation, there may be a change in the proteins, and ultimately in the phenotype, but not always. Much depends on how the manufacture of the particular protein is affected, where and when the mutation occurs, the complexity of the genetic control for the trait and the built-in safeguards against error in the genetic system.

If the trait is governed by the mutual and equal interaction of a dozen gene pairs, the impact of a mutation might be very slight. If the trait is controlled by a single pair, the impact might be much greater. For traits controlled by a major gene pair interacting with several other gene pairs of lesser importance, the effects would depend on which gene had mutated. Whether the mutated gene is recessive or dominant would also be important. It might take several generations for a mutated recessive gene to show up in the phenotype.

Mutations may take place in each and every cell in an organism's body. They will be passed on to the next generation only if they affect the sex cells which produce the gametes which participate directly in the creation of offspring. If the mutation happens in a body, or *somatic*, cell, only cells that derive from the mitotic division of that cell can carry the mutation. If the gene is not involved in the specialized functioning of the affected cell, it won't have a noticeable impact. A mutation for eye color in a muscle cell which has already differentiated and deactivated the portion of genetic program dealing with other organs, such as the eyes, will have no consequences for the organism. Mutations in somatic cells, whether muscle cells, liver cells, brain cells, etc., disappear when the organism dies. They do not become part of the gene pool which serves as the source of genetic variation for the species.

If the mutation happens in the reproductive cells it may be passed on to future offspring. If Queen Victoria's mutation for hemophilia had occurred in a somatic cell, none of her descendants would been afflicted with the disease. The gene would have died with her. A mutation in a sex cell may not necessarily affect the offspring. The mutation may occur in a single gamete and will therefore be passed on only if that particular gamete actually participates in conception. Human semen may contain as many as 50 million sperm cells, so the chances against the mutation carried in a single sperm actually being inherited are very great. Should mutation occur early in embryonic development, all gametes might carry the mutant gene, greatly increasing the chances that the phenotype might be altered in future generations. If the mutation occurs in the sex cells after a person is beyond reproductive age, it would obviously have no significance.

Most mutations tend to be recessive. They are frequently masked by dominant genes for several generations before making themselves apparent in the phenotype. Mutations happen to genotypes, but we can't see a genotype. We can only infer what has happened to the genotype from what we see in the phenotype.

MUTATIONS: HELPFUL, HARMFUL OR NEUTRAL

Perhaps you have seen one of those low-budget horror movies about monsters like giant, mutant bees devouring Cleveland. The effects of mutations are hardly so dramatic and not necessarily so harmful. Since each gene has a definite chemical composition, only a finite number of mutations for each gene are possible. The same limited number of mutations have probably been recurring in each species for millions upon millions of years. By now most of the helpful mutations have been incorporated into the normal genotypes through natural selection.

Mutations cannot produce sudden, drastic changes—like the appearance of giant, man-eating bees—overnight. Many generations are required for them to generalize throughout a species population. Gradual environmental changes necessitated modifications in the eating habits of the ancestors of today's horses in order to survive. To adapt successfully to these new conditions, natural selection for longer teeth was required. The average length of horses' teeth has increased by sixty millimeters (almost 2½ inches), but it took sixty million years for this transformation to be accomplished! Why? Because the genetic program is a delicately balanced instrument. One segment cannot be changed without affecting other parts of the program. In artificial selection, humans can consciously direct and speed up the process of species change. But in the natural environment, the situation is very different. In the case of the horse, nature could only permit an increase of one millimeter every million years and still maintain an overall balance in the horse's genetic program.

It would take millions of years for a strain of giant bees to evolve, and this would happen only if such a change were beneficial, or at least neutral, to survival. Since giant bees would drastically upset the balance of life and the food chain, it is hard to imagine that their development could be anything but harmful. If such a change did not enhance survivability, natural forces would work against its generalization throughout the species population.

Perhaps because of Hollywood, there is a popular belief that mutations are harmful. It's true that many mutations occurring today are less than helpful, but this is because most helpful ones have already been incorporated into the normal genotype. Detrimental mutations tend to be discarded by natural selection, but keep recurring at random again and again. It would be a mistake simply to think of mutations as harmful. Mutations are the source of genetic variability. Without mutations there would have been no evolution, no development of life. The richness of life on Earth, which is a richness of difference, stems ultimately from mutation.

WHAT CAUSES MUTATIONS

Mutations result from a number of causes. In a complex process like gene replication, accidents are always possible. The amazing thing is that so few happen. Some genes are more susceptible to mutation than others. The gene causing hemophilia mutates 20 to 30 times per million gametes produced. In general human genes can be expected to mutate once every 30,000 to 50,000 times that they are duplicated.

Many factors can influence the frequency of mutations. Maternal age is correlated with increased mutation in the egg cells. The older the mother is the greater are the chances that mutation will occur. Certain mutations in sperm cells are related to the age of the father.

In addition to natural accidents, there are certain substances or agents that may be present in the environment that can induce mutations. Known as *mutagens,* these substances may cause a mutation directly, or indirectly, by triggering chemical reactions within the cell that lead to a breakdown in the normal structure of a gene or a group of genes.

Radiation can significantly accelerate the normal mutation rate. When radiation passes through the cell, it releases tremendous energy which can cause chemical alterations within chromosomes. This radiation can come from natural sources like the rays of the sun or radioactive material within the Earth. Or it can come from man-made sources like x-rays.

The mutagenic potential of x-rays was first discovered about fifty years ago in laboratory experiments by Hermann J. Muller. Under undisturbed natural conditions, the average fruit fly (drosophila) gene mutates about once in a million generations. In 1927, Muller found that he could increase this rate by 150 times by exposing the reproductive cells to large doses of x-rays. Today doctors and dentists are very careful in the way they use x-rays. X-rays are helpful in diagnosing internal disorders and broken bones, but these beneficial things have to be weighed against the mutagenic drawbacks. Because mutations in the reproductive cells are likely to affect heredity, doctors are careful to cover your lap with a lead apron when x-raying your teeth, or your chest. Doctors are very reluctant to use x-rays during the early months of pregnancy because a mutation in the embryo or fetus at that stage of development might be included in many cells and possibly cause birth defects. Once the fetus is more fully developed, the impact would be less serious.

In the next chapter we will discuss the nature of DNA, the substance in which the genetic information is coded and examine how mutations actually occur.

CHAPTER TEN

DNA and the Genetic Code

Cells contain many substructures which execute specific tasks necessary for the maintenance of life. These functions are not carried out haphazardly but in a definite predetermined manner based on the interaction between the environment and the genetic program. A system of feedback and signals has been devised by nature to give the genetic program informational input from the environment and various substructures within the cell. This information enables the organism to make needed adjustments and triggers the next appropriate response in the life cycle under the circumstances prevailing at the time.

How this works can be seen in a simple bacteria cell. Like all cells the bacteria cell requires certain substances for survival and growth. The bacteria cell produces some of these internally. Others must be obtained from the environment. The bacteria membrane can distinguish between the different substances that may be present in the environment. It forces those that are needed through and into the cell. It blocks those that are either unnecessary or harmful from entering. If it happens that some of the substances that can be produced internally are present in the environment, the membrane will push them into the cell as well. Simultaneously, the bacteria cell will halt whatever internal process that is manufacturing these substances. When the external supply is used up, the internal process will begin again. This would be impossible without a system of feedback and signals between the components of the cell.

The way this feedback system operates can be compared to the thermostat in your home. When the temperature drops to a certain level, the thermostat gives off a signal which turns on the furnace. The furnace supplies heat and warms up the house. When the temperature reaches another preset level, the thermostat emits another signal which turns off the furnace.

In higher forms of life the interconnection between environment, genetic program and the feedback is even more intricate. The development of a human being from a single cell to an organism containing millions of cells is controlled by the genetic program, heavily influenced by environmental conditions and regulated by a system of feedback. This development follows a definite sequential pattern. It produces very precise arrangements of highly specialized cells which make up the body organs, tissues and nerves. Genetic information is translated into action when needed depending on the development and specialty of the cell. Information on the construction of an eyeball is translated into action before birth, not at ten years of age. As embryonic development proceeds, cells become increasingly differentiated. Unnecessary segments of the genetic program in particular cells are deactivated or turned off. Skin cells don't have to translate the same information as brain cells, but they certainly need to use the part of the code that carries information about pigmentation and gland formation.

Genetic information doesn't act directly on the life processes within cells. It uses proteins, acting as intermediaries, to put its instructions into action. There are many types of proteins that can be used. One important group are the enzymes. Enzymes can speed up or initiate chemical reactions within the cells. Another group are the structural proteins which are used to construct cartilage, bones, cell membranes and other parts of the body. There are also special proteins which activate the translation of segments of the genetic program at the appropriate moment.

Proteins are manufactured in the cell cytoplasm—the material surrounding the nucleus—by linking together molecules of amino acids. There are 20 different amino acids. The exact sequence in which they are linked together determines the protein produced. Different sequential arrangements make different proteins. The genetic program carries precise instructions which guide the cell in arranging amino acids and tell the cell when to terminate a sequence in the building of proteins.

DNA

The chemical substance that carries this blueprint for life is *deoxyribonucleic* acid, or DNA, as it is almost universally referred to. DNA codes the genetic program for all living things. It is found in the nucleus of every living cell from the simple single-cell creature to the complex multicellular organism.

Physically, DNA appears as long double strands. Each strand is actually a chain of units called *nucleotides* linked together. Each nucleotide itself has three distinct parts:

——a sugar called deoxyribose

——a phosphate group

——and one of four chemical bases.

These four bases are the key to the genetic code. They are

adenine (A), cytosine (C), guanine (G), and thymine (T). In much the same way as the twenty-six letters of the alphabet are used to code specific words, the precise arrangement of the nucleotides with their particular bases determines the information being communicated.

A series of three nucleotides stands for a single amino acid. Different combinations of three nucleotides are the coded expression for all twenty amino acids. These combinations also signal where messages start and stop (something like punctuation marks in a sentence). The sequence of nucleotides in a DNA strand codes the sequence of amino acids in the proteins produced in the cytoplasm which direct the living thing's development and survival. (Figure 33)

A single DNA strand is formed by the chemical bonding between adjacent nucleotides in the chain. The double strand is formed by two single strands fitting together like the teeth on a zipper.

Each base in one strand has a mutual attraction for the base at the corresponding location on its matching strand. These pairings are not random. They always follow a pattern. Guanine (G) always pairs with cytosine (C) and adenine (A) with thymine (T).

DNA strands never lie flat in the nucleus. They twist themselves into a rather strange shape called a *double helix*. If you've ever climbed the long spiral staircase at the Statue of Liberty, you have a first hand acquaintance with a double helix. Think of the handrails as the strands of DNA and the steps between them as the attraction bonds between the bases in each strand. (Figure 30)

In Chapter Three we compared the cell to a computerized factory. We pointed out that the cell is very different from man-made computers because there is no programmer around who can control it. Living things are self-acting and self-reproducing. Because they are so special, living things

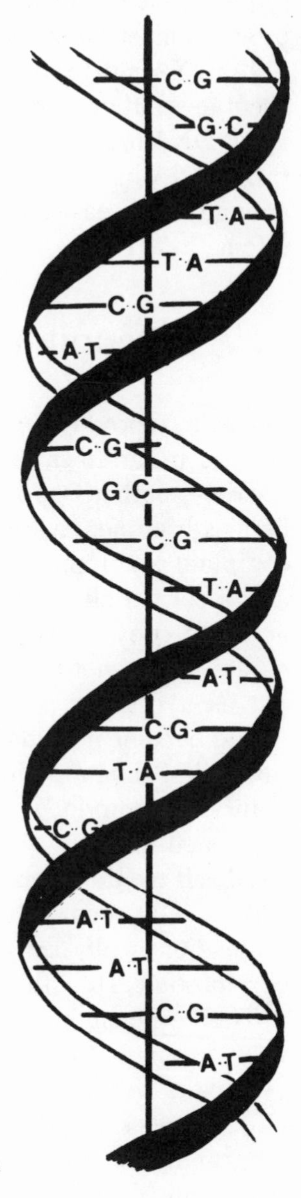

Fig. 30 Double helix structure of DNA

require that the DNA which codes the biological computer program meet three very important requirements:

1) DNA must reproduce itself with sufficient stability and accuracy so that no information gets lost.

2) DNA must also have a potential for change.

3) DNA has to carry the coded message in a form simple enough to be easily and reliably understood and translated into action.

REPRODUCTION OF DNA

Accurate reproduction of DNA is an absolute necessity. The genetic program is duplicated each and every time a cell divides—perhaps as many as 50 billion times in a human being's lifetime. Every cell except the gametes must carry a complete set of genetic information. The gametes carry only a half set. In all cases, it's essential that the cell receive an accurate copy of its genetic material. If replication is inaccurate, the genetic program may degenerate. What started out as a definite, sequentially ordered set of information would become nothing but a bunch of random nonsense. The survival of the individual organism would be jeopardized. Its ability to reproduce would be impaired. If the same thing happened in many species members, the survival of the species itself would be endangered.

As a cell prepares to divide, each gene produces an exact replica of itself, constructing a brand-new copy of each chromosome. How this happens is a fascinating process.

The key element is the fact that the bases in the double strands pair with each other in a definite pattern. Guanine (G) pairs with cytosine (C) and adenine (A) with thymine (T). If you know the sequence of bases in one DNA strand you can figure out the sequence of basis in the corresponding strand quite easily. If one strand is C-C-C-A-A-A . . . , the corresponding strand would be G-G-G-T-T-T . . . , because G always pairs with C and T always pairs with A.

The two DNA strands begin to unzip. Each reconstructs its corresponding strand from free-floating nucleotides in the nucleus. There are many such molecules floating around in the nucleus randomly bumping into each other, into the nuclear membrane and other molecules in the nucleus. According to the laws of physics, we know that millions of such collisions take place in a split second.

Suppose that a C-C-C-A-A-A . . . strand has just unzipped from its corresponding strand. And suppose that in one of the millions of collisions between free-floating nucleotides and other molecules in that split second, a G nucleotide bumps into the first C nucleotide in our strand. Since these two nucleotides have a mutual attraction, it aligns itself at this location. An enzyme acts to chemically fuse this new nucleotide to the strand that is being built up. Thus, this G nucleotide stays put. Now let's suppose another G molecule comes along and collides with the next C in the chain. It too forms a bond. Nucleotide by nucleotide, our strand is zipping itself back up again by creating its corresponding strand.

The other original strand, the G-G-G-T-T-T, is doing the same thing at the same time. When they're done, two identical double strands of DNA will have been constructed. (Figure 31)

DNA AND CHANGE

It's important that the self-reproduction of the genetic material be accurate. But it's also important that there be a possibility for change. If DNA were totally resistant to change, there would be no mutations, no genetic variation, no evolution. Life would still be stuck where it was one and a

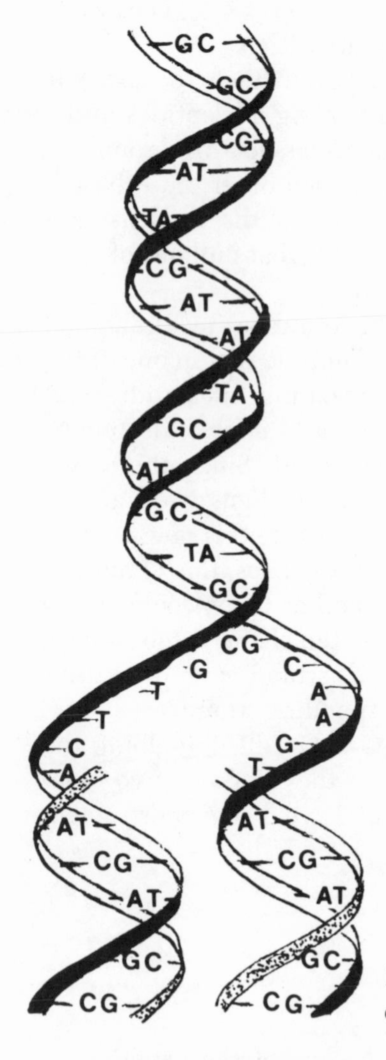

Fig. 31 Replication of the double helix

half billion years ago as a tiny blob of protoplasm. Change within reasonable limits must be possible.

How does DNA allow for change? Let's go back to our unzipping C-C-C-A-A-A . . . strand again. Suppose everything is proceeding nicely. First one guanine and then a second collide with the first two cytosines. But then there's a slight error. Millions of nucleotide collisions are occurring every second. By chance, an adenine, instead of a guanine happens to be the first nucleotide to bump into the next cytosine in the strand. Though they're not normal partners, these two nucleotides are attracted somewhat. It's not as strong as the attraction between cytosine and guanine, but they do pull toward each other. Chances are that if another guanine collides immediately it will bump the intruder adenine away, take its place and restore the correct sequence. But it's also possible that the next nucleotide to come along might be a thymine, which has a strong attraction for the next nucleotide in the original chain, as adenine. The chain building process would continue. The mistake would be sealed in. When duplication is completed, the two double-stranded DNA chains would not be exact replicas of each other. A mutation would have occurred. (Figure 32)

One of the daughter cells produced by this cell division will have a DNA chain identical to the mother cell's. The other cell will have a slightly altered chain. The third pair will have a C-A pairing instead of a C-G. But this is not a stable arrangement. When this cell divides, each nucleotide will tend to attract its normal partner. The C will join with a G, restoring the normal sequence found in the original mother cell. And the A will join with a T, and form a stabilized but altered sequence as compared to the original mother cell. The third pair will be A-T instead of C-G. This alteration will be passed on to all descendants of the cell.

We have already mentioned that the impact of mutations can vary a great deal. Even if the altered information relates to a particular cell's normal activity, the repetition and pattern of the DNA code gives the cell a built-in safeguard against mistakes in DNA duplication. It allows a tolerance for error, without changing the basic message in the code.

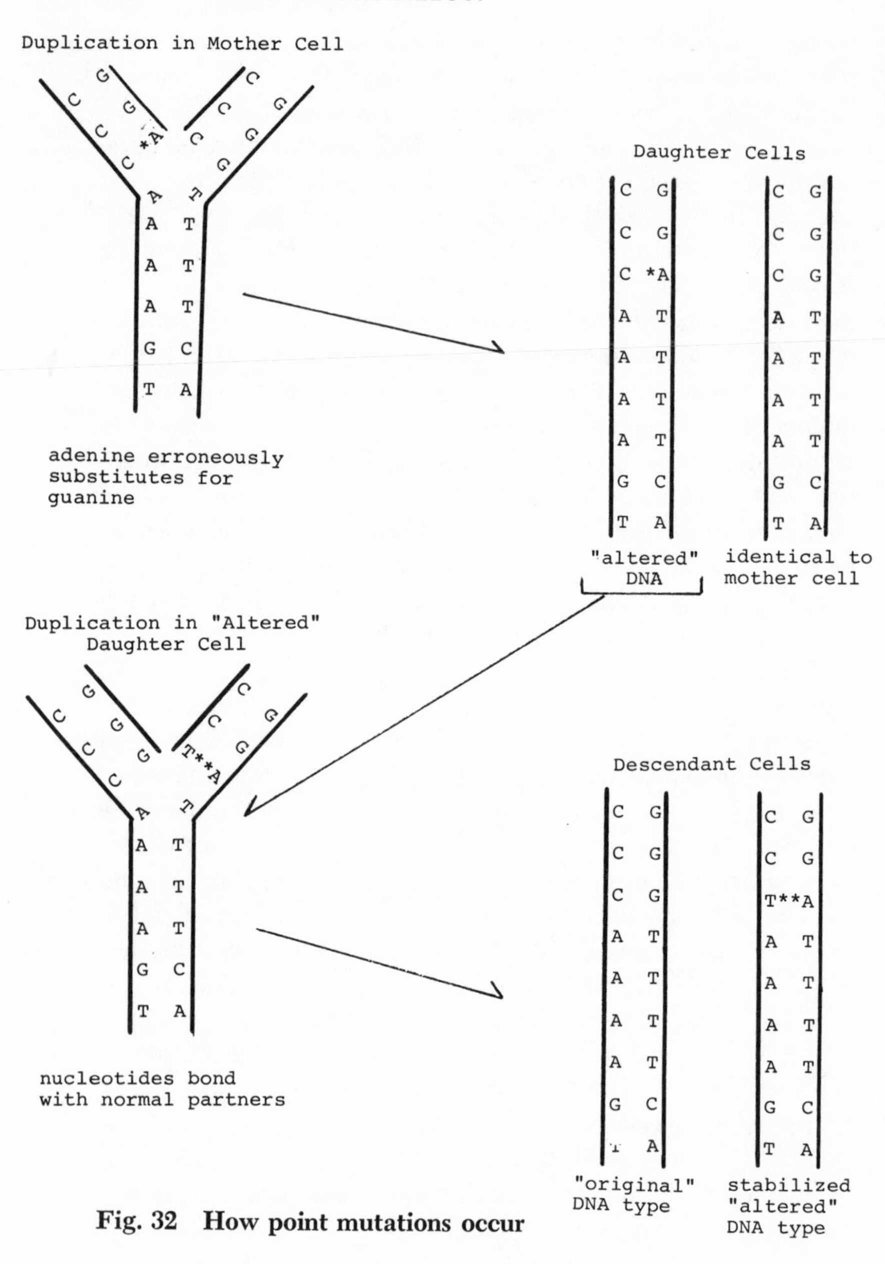

Fig. 32 How point mutations occur

The significance of an error depends on how much and in what way it disrupts the normal pattern.

This ability to communicate information successfully despite occasional errors is true for other means of communication as well. Take the sentence, "He shut the boy in the room." The information conveyed by this sentence is clear enough. Suppose we tried to reproduce this sentence by typing an exact copy of it on a piece of paper. It would be quite possible for even an expert typist to make a typographical error, by substituting an inappropriate letter for one contained in the sentence. How seriously such a typographical error might affect the sentence depends on how it alters the pattern of letters making up the words. If an a were accidently substituted for the letter e in the word the, the sentence would read, "He shut tha boy in the room." No one would have much trouble figuring out the message of the sentence. Someone might wonder if the author meant the boy or that boy, but the basic thrust of the sentence would still be clear.

But suppose the typographical error was the substitution of an o for the u in the word shut. The sentence would then read, "He shot the boy in the room." And that would give a totally different message!

In either case, only one letter was altered. But where it occurred and how it changed the pattern was crucial. It's the same for changes in DNA replication.

Some amino acids can be coded in more than one way. And it's therefore quite possible that a base substitution, while changing the arrangement of a nucleotide triplet, may actually still communicate the same information.

DNA TRANSLATION

The Bureau of Engraving and Printing is very careful in what it does with the master plates for American paper money. These plates, engraved by skilled craftsworkers, are kept under tight security. They are never directly used to print money. They are used only to make printing plates which are put on the printing presses to make the money that eventually goes into circulation.

In a similar way, the cell takes great care to protect its master plan. The DNA-coded genetic program never leaves the safety and security of the cell nucleus. In order to get its instructions out into the cytoplasm where the proteins are synthesized, the DNA needs a messenger, and a way to decode the message. This involves a two-part process: transcription and translation.

The DNA in the nucleus transcribes a copy of itself into a substance called *ribonucleic acid*, or RNA. Like DNA, RNA is composed of nucleotides linked together in strands. There are, however, four important differences between RNA and DNA:

1. The sugar involved is ribose.
2. RNA is found only in single—not double—strands.
3. Although it also uses four bases to code the genetic information, RNA doesn't contain thymine. Thymine is replaced by uracil (U). Wherever thymine appears in DNA, uracil is substituted in RNA.
4. RNA is not as stable as DNA—it degenerates shortly after doing its job in the translation process.

Transcription of the genetic code into the RNA is very similar to DNA replication. A strand of DNA builds up a strand of RNA containing corresponding bases (with the exception of uracil substituting for thymine). Our C-C-C-A-A-A DNA strand would therefore produce a G-G-G-U-U-U RNA strand.

Since this RNA strand is sent out from the nucleus to the cytoplasm it is called *messenger RNA (mRNA)*. The mRNA carries an accurate copy of the information in the DNA. As in DNA, each triplet set of nucleotides specifies a specific amino acid. The arrangement of these triplets determines how the amino acids will be put together and which protein will be manufactured. (Figure 33) The mRNA passes through the nuclear membrane and proceeds to the ribo-

Fig. 33 The genetic code

TRIPLET IN DNA	TRIPLET IN mRNA	AMINO ACID CODED	TRIPLET IN DNA	TRIPLET IN mRNA	AMINO ACID CODED
ATA	UAU	Tyrosine	AAA	UUU	Phenylalanine
ATG	UAC		AAG	UUC	
ATT	UAA	Termination	AAT	UUA	Leucine
ATC	UAG	of chain	AAC	UUG	
GTA	CAU	Histadene	GAA	CUU	Leucine
GTG	CAC		GAG	CUC	
GTT	CAA	Glycine	GAT	CUA	
GTC	CAG		GAC	CUG	
TTA	AAU	Asparagine	TAA	AAU	Isoleucine
TTG	AAC		TAG	AUC	
TTT	AAA	Lysine	TAT	AUA	
TTC	AAG		TAC	AUG	Methionine
CTA	GAU	Aspartic acid	CAA	GUU	Valine
CTG	GAC		CAG	GUC	
CTT	GAA	Glutamic acid	CAT	GUA	
CTC	GAG		CAC	GUG	
ACA	UGU	Cystine or	AGA	UCU	Serine
ACG	UGC	cystein	AGG	UCC	
ACT	UGA	Termination	AGT	UCA	
		of chain	AGC	UCG	
ACC	UGG	Tryptophan			
GCA	CGU	Arginine	GGA	CCU	Proline
GCG	CGC		GGG	CCC	
GCT	CGA		GGT	CCA	
GCC	CGG		GGC	CCG	
TCA	AGU	Serine	TGA	ACU	Threonine
TCG	AGC		TGG	ACC	
TCT	AGA	Arginine	TGT	ACA	
TCC	AGG		TGC	AGG	
CCA	GGU	Glycine	CGA	GCU	Alanine
CCG	GGC		CGG	GGC	
CCT	GGA		CGT	GCA	
CCC	GGG		CGC	GCG	

somes in the walls of the endoplasmic reticulum which we mentioned in Chapter Three. It is here that the message is translated.

The ribosomes are yo-yo-like structures made up predominantly of another type of RNA called ribosomal RNA. The ribosomes are comparable to the printing presses used to make money. The mRNA strings itself through the ribosomes. Still another type of RNA is present in the cytoplasm to help complete the translation process. This is the *transfer RNA (tRNA)*, which is also produced in the nucleus and passed through into the cytoplasm. tRNA bounces around the cytoplasm and bonds with different amino acids. The tRNA guides the appropriate amino acid to the translation site at the ribosome in the right sequence. The amino acids are bonded together to form the specific protein directed by the genetic code transcribed in the mRNA.

The whole process from transcription of the genetic message into the mRNA to the production of the finished protein takes less than two minutes. It happens simultaneously at millions of ribosomes in each cell. Depending upon the cell specialty, transcription and translation may be repeated over and over again. The cell is indeed a very busy place.

The information coded in the DNA is an absolute prerequisite for life, but by itself it is not enough. It requires the right environmental conditions. It's very much like building a house. You need architectural plans to know what to do. But no matter how good your blueprints, you can't build a quality house without bricks and mortar and skilled bricklayers. The genetic blueprint is no different. Without the right material being present in the environment, the genetic program can't be put into action successfully.

CHAPTER ELEVEN

Genetics and the Environment

In the beginning of the twentieth century, an American scientist suggested that there was a gene which made the young men of fishing villages go to sea at the age of 18, 19 or 20. In ancient times, the Egyptian King Psammeticos ordered two young children to be raised in complete isolation from other human beings in order to discover the language that humans would instinctively communicate in. In 1947, C. D. Darlington, a British writer, argued that people who carry the gene for O blood type are best able to pronounce the th sound of the English language.

All three of these ideas reflect a long-standing belief in instinctual behavior—the idea that people inherit behavior patterns that they don't have to learn. What's wrong with these theories isn't too hard to see.

Young men in the fishing villages of the late nineteenth and early twentieth centuries were raised by families whose entire existence revolved around the sea. As youngsters they spent their time on the wharves and in the boats. Many fishing villages had no other industries that could give them jobs when they reached young adulthood. They could move away and seek a job elsewhere, or they could go to sea. For them, going to sea was no more genetically determined than going into the mines was for a young man raised in a coal-mining town in West Virginia, or going into a steel mill was for a young man in a town where the only place to get a job was the steel mill. The environment they grew up in was the critical factor, not their genes.

As for the ability to speak language, children learn to speak whatever language they hear as they grow up. If they're raised in isolation they won't get beyond gurgling sounds, and what's worse, they will be mentally, emotionally and socially retarded in their development. Children need to learn from the people around them. This requirement applies to all aspects of their behavior, including language. Human beings are social beings. Cultural skills are not genetically determined, but learned.

When a female laboratory rat gives birth to a litter of pups, it will normally follow a special behavior pattern. It will lick and clean its babies. It will protect them. It will feed them. It will do this without ever having read Dr. Spock or being taught to do so. Kittens as well as all other young animals receive similar treatment from their mothers. People call this the *maternal instinct*. They say that these female animals have inherited a way to behave as mothers. They say that this is an inherited behavior pattern which is necessary to make sure that the babies survive and continue the species for still another generation.

The argument for instinctive behavior in animals seems more convincing than it does for humans. How humans learn things and are influenced by the world around them seems easier, perhaps, to understand. But the overwhelming evidence is that there is no such thing as innate behavior, including maternal instinct. What these mother animals do when they have babies is environmentally conditioned, just as going to sea was for the young men of fishing villages.

This has been proved in laboratory experiments. If a female rat is raised with a large rubber collar around her neck that interferes with her ability to touch the rest of her body with her tongue, her "maternal instinct" disappears. When she has pups, her behavior is totally inappropriate. She will not care for them, clean them or protect them. She may actually kill them and eat them!

What the mother rat inherits is not maternal behavior, but the potential to develop acceptable maternal behavior. She is genetically programmed to learn about her own body. She keeps herself clean and well-groomed by licking herself. In so doing she learns about body odors and tastes and feelings. This experience gives her preparation for following a similar pattern of behavior when she has babies. When the rubber collar is placed around her neck as a young rat, she is denied the chance to clean and groom herself properly. She therefore doesn't experience her own body in the normal way and is unprepared to be an adequate mother to her babies.

What we are at any given moment in our lives—how we look, how we behave—depends on the interaction between a genetic program and an environment. We do not inherit traits or behavior patterns. The genetic program does not automatically determine such things. It merely gives us the potential to survive and develop in a given direction in the environment we live in. How far we go in fulfilling this potential is a product of the complex interaction between environment and genotype. This is true for all living things. The interaction we are talking about is so intricate and so interwoven, that people have often had difficulty telling what's genetic and what's environmental.

A classic example is the breed of laboratory mice which

was thought to have a genetic predisposition to a high cancer frequency. These mice were born healthy but developed cancer as they grew older in large numbers. An experiment was conducted using this breed and another breed which almost never developed cancer. Litters from each breed were exchanged at birth even before they had a chance to suckle at their mother. An amazing thing happened. The cancer rates for these offspring were reversed. Babies from the high-cancer-rate breed wound up having a low incidence. And the babies from the low-cancer-rate group wound up with a high incidence. The cancer was not genetically caused at all. It was the result of a cancer-causing substance present in the milk of the mother in the high-cancer breed. Any baby fed on this milk—whatever its genetic makeup—was in danger of developing cancer as it grew older!

The word environment means more than clean rivers, clean air and thriving wildlife. To the geneticist, environment refers to everything outside the individual that impinges upon and influences its growth and life. It includes obvious things like temperature, atmospheric conditions and the availability of food. But it also includes less obvious factors in different cases like what your mother ate when she was pregnant with you, or how often your father smiled at you when you were an infant, or whether a female rat had the experience of licking and cleaning herself before she had pups.

The genetic program has developed over millions of years to permit life within a range of environmental conditions. Humans can survive in climates ranging from the tropical heat of tropical Africa to the cold of Greenland. But the genetic program has not been designed to function successfully in the sizzling heat of the planet Mercury where the temperature may reach between 500 and 770 degrees Fahrenheit—hot enough to melt lead and tin. When astronauts walk on the moon they carry a simulated version of the Earth's atmosphere with them in pressurized space suits because they are not genetically equipped to survive in an atmosphere as thin as the moon's. They are equipped with the mental capabilities, however, that allow them to figure out a way to get to the moon and to create a simulated atmosphere in a space suit.

The genetic program of mammals requires that they breathe air. Whales and dolphins must come to the surface to breathe or they drown, no matter how much they resemble fish. The genetic program of fish denies the possibility of survival outside of water.

All living things need food because their biological computer programs do not give them the power to synthesize all the materials necessary for life. They must get some of them at least from the world outside themselves. Whether these necessary foods are available in the particular environment, in what quality and what quantity, has a bearing on how well a living thing grows and develops, or whether it starves to death.

Environmental conditions set the framework for life and death. No air, we die. No food, we die. Temperature too hot, we die. Within the range of environmental conditions in which the genetic program is designed to operate, the interaction between genetics and environment is still important. It affects the way we develop, the way we behave and the way we appear. The genetic program gives everyone the capacity to develop a human physical appearance within normal limits. There is plenty of variation between individual people—no two look exactly alike. But there is variation only within certain limits. There are no people with six legs or three heads, but some individuals, for instance, do inherit a potential to be heavier or thinner than others.

Whether a person is fat or thin is not simply genetic. There may be a significant genetic component, but environmental conditions are also involved. When people "watch their diets" to lose weight, they are trying to regulate the environmental influences on their genetic program. A person may be fat or chubby from eating too much and/or not exercising enough. "Fatness" may sometimes seem to run in families. A father and mother may be fat, and their children may be fat. But is it genetic or is it environmental? Members of a family share more than their genes. They also share a common environment where maybe too much food is placed on the table, where too many fattening foods are served, and where exercise is discouraged. In certain parts of the world there is an inadequate supply or distribution of foodstuff, and famine threatens to destroy human lives by the thousands. Perhaps you have seen pictures of starving children with the swollen, distended bellies that result. They didn't inherit this physical appearance. No one should have any problem seeing the relationship of the environment to these physical characteristics.

The genetic program gives humans the same basic set of muscles. But the muscle tone and appearance of a man who works out in a gymnasium every day is going to be very different from that of a man who sits behind a desk all day and never exercises. The type of exercise you do is important, too. The muscle tone of a basketball player will differ from that of a weight lifter because of the drastic differences in what they do with their muscles and how this activity affects the way their muscles develop.

In the coal-mining regions of West Virginia and other parts of the Appalachians, there are families in which the grandfathers, sons and grandsons all suffer from a condition known as the black lung disease. The people in these rugged isolated mountain areas have kept to themselves over the years. For generations they have married amongst themselves. But black lung disease is not genetically transmitted. It is caused by inhaling coal dust while working in the mines.

The United States government officially recognizes any person who has put in thirty years in the mines as a victim of the disease. With or without physical examination, anybody who has worked in that environment for three decades is eligible for a special disability pension.

There are many other occupationally linked diseases and conditions. Bakers suffer lung problems from inhaling flour over the years. Recently, the newspapers have carried accounts about the dangers of asbestos. Workers who were exposed to this construction material while working on World War II navy vessels are in jeopardy. In addition, there are constant reports about the cancer-causing properties of various food additives and drugs. The problem here is that the genetic program has not prepared our bodies to adjust to many of these newly developed environmental stresses. The industrial period is very recent, just seconds in the evolutionary clock. Perhaps as many as ten thousand chemicals have been invented in the last century and added to our foods and industrial processes. All of these substances represent elements in the environment, which evolution (which takes millions of years to get its job done) hasn't had time to develop safeguards against.

Many of the things we have talked about so far are gross environmental conditions. But there are far more subtle ways in which the genetic program and environment interact to make you what you are. We will discuss these factors in later chapters in more detail.

NATURE VERSUS NURTURE

The more we learn about genetics, the more important the environment seems to be. Sometimes it is difficult to understand how the genes manage to assert themselves over the complications imposed by the environment. Plants have green leaves. What could be a more obvious example of genetic determination? But it's not that simple. Plants have green leaves because they have chloroplasts which manufacture chlorophyl, which is green. To do this, the chloroplasts must have exposure to light. Laboratory experiments have been done in which plants have been carefully raised in darkness, without access to light. The leaves were white, or colorless. This illustrates what we've been saying about the inheritance of potential. Plants inherit the potential to have green leaves under the correct environmental conditions.

Which is more important—the heredity or environment? Historically, this has been the subject of the great "nature vs. nurture" debate. We have already referred to the widespread belief in instinctual behavior. In the nineteenth century, the tendency was to belittle environmental influences and stress biological inheritance, especially in explaining human behavior.

A well-known study was made of important family histories which claimed to prove that outstanding achievement ran in families. The Bachs were cited for producing fine musicians. The ancient Roman family, the Scipios, was cited for its generals and orators. These achievements were said to be the result of biological inheritance. Environmental factors, such as wealth and family tradition were discounted. The man who did this study chose to overlook a fact that actually makes a strong argument for environmental influence. One of the Scipio generals was an adopted son, who shared only family tradition and not biological inheritance with the rest of the Scipios.

The underestimation of environmental influences persisted into the twentieth century. The supposed genetic causes for young men going to sea is just one example. In 1911, Davenport published an important study which traced the accomplishments of New Haven's Tuttles-Edwards family and Virginia's Lee family. The Tuttles-Edwards family included two presidents and one vice president of the United States, and six college presidents. The Lees produced a long list of political figures and military generals. Davenport then contrasted these achievement-oriented families with two other families, the Jukes and Kallikaks who had given rise to generations of prostitutes, thieves and drunkards. The differences between these two sets of families were attributed to genetics. Environmental conditions were belittled, despite the fact that it's hard to visualize how a person born with the capabilities of becoming a college president who was raised by prostitutes and thieves could ever live up to his potential, especially in the eighteenth and nineteenth centuries. He would have found many factors stacked against him.

This approach which held that a person's profession was a result of his/her genes was carried to an extreme in 1912 by W. Bateson. Bateson wrote that "acrobats, actors, artists, clergy, farmers, laborers, lawyers, mechanics, musicians, poets, sailors, men of science, servants, soldiers and tradesmen" were different genetic "types and strains." He thought that the social differences between people were genetically determined, and expressed surprise that it was still possible for people from such diverse walks of life as farmers and artists to mate and produce healthy offspring. To him, they were practically different species!

These ideas die hard. In the beginning of this chapter we mentioned Darlington's thoughts on genotype and the ability to pronounce the th sound. In 1953, Darlington complained that England's agricultural problems were caused by "the genetic unfitness of a large part of the tenant farmers." And in the United States even today there are serious theories put forth which claim that blacks are genetically inferior to whites. These theories dismiss the complex social problems we face in society today as simple biological conditions.

The idea that behavior is inherited is wrong. It is just as

wrong as the idea that we inherit traits ready-made. The only thing that any living being inherits from its parents is a set of genetic information coded in DNA. Everything else is a product of the interaction between this genetic program and the world around it. Let's examine one critical area of human behavior—language—to see how this interaction works.

HUMAN LANGUAGE

Many writers have pointed out that language is a key human behavior that separates us from the lower animals. Humans have the ability to conceptualize things they see in the environment, to invent symbols for them and to express them in language. Language is found in all human populations. The content and complexity of language vary from culture to culture, but even the most primitive society communicates by means of the spoken word. Those individuals who are physically disabled from speaking or hearing can still abstract information from the environment, symbolize it and communicate it. They merely translate the information into another form—sign language or writing.

How complicated a language is depends upon the social environment in which it's used. This includes how many words it has and what they are. Primitive tribes in isolated jungles may not have need for a word like electricity. They have no experience with it; it is not a part of their social experience. Ancient Hebrew and Latin never had a word for television, and neither did modern English until the TV was invented. Language is not inherited; it is socially conditioned and develops with society. Even the words that make it up are adapted to express what needs to be said about the environment.

Dr. James King has pointed out that human babies follow the same pattern in learning how to speak in every society. In the first year they begin to experiment with random sounds. Soon they put these sounds together to make words. They then begin to speak in phrases. By age four, the child can usually speak in full grammatical sentences.

As King Psammeticos discovered, there is no innate human language. Language is a social skill. The particular language a child learns is completely decided by the environment he/she is raised in. There is nothing genetic about it. Suppose a newborn German boy were adopted by a Japanese family and taken to live in Tokyo. He would learn to speak and think in Japanese with the same facility that he would have learned German. He would speak Japanese without accent or hesitation. And he wouldn't know a single word of German.

We inherit genetic information which enables us to develop and speak a language. The genetic program has given humans a marvelous neurological system that gives us a brain that can act like a computer. Not only can our brain-computer think, and symbolize, and remember, but it also gives us the potential to control behavior. We also have certain physiological capabilities which can work in concert with our neurological system and allow us to manipulate our voices to make the sounds of language.

Many parts of the genetic program are activated sequentially for specific, limited periods. The part of the program controlling language is one of them. Try as hard as you can but you can't get a two-month-old baby to recite Shakespeare. It is just too early for speech to be possible. Up to the age of puberty, a youngster has practically no problem learning a second or even third language. King points out that children raised in homes where the parents speak two languages, usually learn both quite easily. And they can differentiate between the two without confusion. They can use the right grammatical rules for each language, and they may even speak both without accent.

With the onset of puberty, the genetic program appears to deactivate part of this remarkable ability to acquire languages. Learning a new language may become torturous. High school students learn their foreign languages by rote. They have to write their vocabulary words over and over again in order to learn them. Speaking without an accent is difficult, if not impossible.

The establishment of language skills is very far along in the development process. But the environmental impact on our development as human beings begins long before birth.

CHAPTER TWELVE

Prenatal Development and the Environment

Thalidomide was a new wonder drug that was put on the market in a number of European countries in 1960. As an antinausea pill and sedative, it was hailed as the cure-all for the discomforts of early pregnancy. No more would pregnant women have to be bothered by morning sickness. The drug was never put on the market in the United States because of the efforts of Dr. Frances O. Kelsey, an official at the Food and Drug Administration. Dr. Kelsey felt that the drug's safety hadn't really been proven. She resisted intense industry pressure to allow thalidomide on the market in this country.

The horrifying side-effects of thalidomide were soon obvious. Within two years, some seven thousand children were born in Europe with terrible birth defects which were caused by the simple fact that their mothers had taken the thalidomide pill during the early months of pregnancy. In 1962 President John F. Kennedy awarded Dr. Kelsey a gold medal for distinguished civilian service to her country.

The thalidomide story emphasizes the fact that the environment affects life even before birth. It points out the importance of the prenatal, maternal environment and the delicate nature of the internal environment within the developing organism itself.

It would be wrong to think of the environment as something purely external. Complex organisms are built of units called cells. Each cell has its own immediate environment made up of the cells around it. These cells interact with and influence each other. This is the internal environment. The way cells influence and signal each other is chemical in nature. The precise workings of this process still mystify scientists. But the crucial importance of this chemical signaling was demonstrated by thalidomide. Thalidomide caused birth defects because it interfered with the normal chemical signals between cells in the embryo. When the mother swallowed the pill, she introduced a strange chemical from the external environment into her body. It passed through her circulatory system and through the placenta into the developing child in her uterus. This strange chemical fooled the genetic program. It emitted false signals which disrupted the normal development of the embryo in a tragic way.

Since the genetic program operates sequentially, the exact type of malformation caused by thalidomide depended upon when the mother swallowed the pill. It did its most damage between the 35th and 50th day of pregnancy. Early in this period it caused deformities of the ear. Slightly later it caused a severe shortening of the arms so that they resembled a seal's flippers more than a human's arms. Later in this period it caused defects of the legs. By the end of this period its impact was restricted to deformities of the thumbs.

The significance of the internal environment begins practically at the moment of conception. The first environmental influence a new organism comes in contact with is the cytoplasm of the egg cell. The food supply concentrated in this cytoplasm is essential for the genetic program to begin its job. Nature clearly had its reasons for concentrating the cytoplasm in a single egg cell during meiosis. The nucleus of the fertilized egg cell is a unique combination of genetic information contributed by both parents. The food material in the cytoplasm comes solely from the mother. It is not really a product of the new organism created by fertilization. It can be considered an informational input from outside the new organism. It is in fact a very specially arranged package of nutritional information designed to meet the needs of the species' own genetic program. Egg cytoplasm differs from species to species, and is not interchangeable. Experiments have shown that if you transplant the *nucleus* of one frog egg cell into the *cytoplasm* of another, the embryo will develop normally. The same is true of salamanders. But, if you transplant the nucleus of a frog into the cytoplasm of a salamander, or vice versa, the embryo will die. If the new organism doesn't have the proper food supply available at the very first moment, life is doomed.

In animals, the first step in putting the genetic program into action after fertilization is a series of quickly repeated cell divisions. These continue until there is a mass of about 5,000 to 10,000 cells, shaped in either a hollow sphere or ellipse depending upon the species. These cells are all exactly alike. No *apparent* differentiation has yet occurred.

Once it reaches the 5,000 to 10,000 cell level, division stops temporarily. The next stage is a process called gastrulation. The cell mass rearranges itself into three distinct layers:

——the ectoderm, which will develop into the skin

——the mesoderm, which will become the skeleton and muscles

——the endoderm, which will become the internal organs.

Even as it develops, the living thing is a self-regulating thing. There is no master chef watching the pots, deciding when to add the next ingredient and when to stir. A living thing cooks itself. The immediate environment of each cell layer is its neighboring layer. The three layers need each other's presence to develop normally. They give each other chemical feedback and signals which trigger the next steps in development. These signals must be genetically controlled. There is no other possible explanation.

Each cell layer induces continued change and growth in the layer next to it. The first is called the primary induction. The mesoderm and ectoderm interact to start the organization of the embryonic body. The ectoderm begins the nervous system with the rudiments of the brain at one end trailing into a spinal cord at the other end. The mesoderm takes the first steps in forming what will ultimately become the spinal column. What's important is that the mesoderm and ectoderm need each other's presence to do their jobs. Experiments with amphibians have been done in which a piece of mesoderm was transplanted from one embryo to another. If placed in contact with a piece of ectoderm that wouldn't normally come in direct contact with the mesoderm, an amazing thing happened. A new primary induction took place, and a new embryo developed as a Siamese twin. The mere presence of the strange mesoderm triggered a reaction in the ectoderm which produced a whole new nervous system and a whole new spinal cord, and a whole new individual. Both embryos developed normally afterward but were born joined at the belly.

There are many other examples of induction we could cite. Soon after the formation of the brain, two projections grow out from the forebrain. At the end of these outgrowths are the optic cups which will eventually hold the eyes. When an optic cup comes directly in contact with the skin in its neighboring layer of cells, it induces the formation of the eye lens. If an animal's eye cup is surgically removed in a laboratory experiment, no lens will develop. If the optic cup is transplanted elsewhere and put in contact with skin cells that might normally make up a leg, it will again induce the formation of a lens. There is something about the cells of the optic cup that signals the skin cells what to do and how to organize themselves.

After the lens has formed, the skin covering the eye continues to develop differently than other skin cells. It never develops any glands. It loses its pigmentation and becomes transparent. Eventually it becomes the cornea of the eye. This is another example of induction. It will happen only in the presence of the eye. If the embryonic eye is removed in an experiment with a laboratory animal, normal skin will cover the eyeless area. If the eye is transplanted and placed in contact with a strange piece of skin, that skin will develop into a cornea. If a piece of skin is transplanted from another part of the body and placed over the eye, it too will lose its glands and pigmentation and become transparent. It will become a cornea even though it might normally have been destined to be the skin on the back or the leg.

The time when a specific induction will occur is controlled by the sequential functioning of the genetic program. The varying impact of thalidomide illustrated when specific organs were being induced.

MATERNAL ENVIRONMENT

Even before conception there are elements in the maternal environment which have an important bearing on whether a particular pregnancy will successfully produce a normal healthy child. The age of the mother can be quite critical. Men constantly produce their sperm cells. But when a woman is born, she carries in an immature form all the egg cells her body will ever have—about 300 in number. Each month one of these eggs matures. If it isn't fertilized by a sperm cell, this egg is shed by the body at the time of the menstrual period. The older a woman gets, the older her egg cells get. There is more time and more chance for something to go wrong with the egg cells as a woman ages: this might result in some type of birth defect in the offspring. In fact, the incidence of certain types of birth defects caused by chromosome abnormalities is much higher when the mother is forty than when the mother is twenty.

For nine months the uterus is the environment of the developing child. The cells in the embryo interact and signal each other as to how to develop and what to do next, but the conditions in the uterus set the framework for the success of overall development. The embryo also interacts with, and is influenced by, this "external" environment within the uterus. If things go well, development is normal. If things go wrong, normal development can be interfered with, or it can stop altogether. The developing child can die even before it is born.

Nature has designed the uterine environment to protect the embryo and fetus as much as possible. (The term "embryo" applies to the developing offspring up until the eighth week of pregnancy; thereafter the term "fetus" applies.) In the beginning of pregnancy, an amniotic sac forms to surround and shelter the baby in the uterus. The sac fills with a fluid secreted by the mother's body. Eventually, fetal excretions are added to the amniotic fluid as well. The embryo floats freely in the amniotic fluid, attached to its mother only by the umbilical cord. This allows normal and symmetrical growth to take place. If the baby were touching the mother's body it might negatively affect bone structure and skin development.

The amniotic fluid also prevents the baby from being hurt by sudden jolts experienced by the mother. If the mother trips or falls, the baby sloshes around in the fluid, rather than being smashed against the mother's body. The fluid also maintains a constant body temperature and allows the baby to move its muscles easily and to exercise, which is an important part of normal development.

The uterus does a good job in protecting the baby from physical blows from the external environment, but there are other ways that the outside world can disrupt normal development. Exposure to harmful radiation or chemical substances (like thalidomide) ingested by the mother during pregnancy may cause malformations at birth. Certain virus infections can also cause congenital birth defects. Substances which cause such birth defects are known as *teratogens* ("monster" inducers). Teratogens differ from mutagens in that they do not change the genetic material; they chemically disrupt the normal path of development by giving false signals. A child born with defects induced by teratogens could have children who would not be affected.

The embryo is especially vulnerable to teratogens in the early stages of pregnancy when differentiation of cells is first beginning. Problems caused by teratogens within the first two weeks of fertilization usually result in early death and spontaneous abortion. From the 15th to 60th day of pregnancy, teratogens may still cause death, but physical abnormalities are more likely. When exposure occurs early in pregnancy the deformities are likely to be severe. Later exposure may result in only minor abnormalities.

Radiation can damage the central nervous system and the eyes, and cause mental retardation in the baby. Radiation may also cause mutations as we noted in an earlier chapter.

The rubella virus which causes German measles can cause cataracts, deafness and ear defects. Certain hormone treatments can cause masculinization of baby girls (a distortion of female genitalia). Certain anti-tumor drugs have been found to cause skeletal and central nervous system defects and stunted growth as well. The antibiotic drug tetracycline may cause tooth defects if taken by the mother during the last six months of pregnancy. If taken in extremely large doses, it has been known to cause cataracts. (Penicillin is apparently quite safe.) Because of these teratogenic effects of drugs, physicians are extremely careful in what they prescribe for pregnant patients. Sometimes, however, even the woman herself is not yet aware that she is pregnant in the first few weeks, and potentially harmful drugs may be taken accidentally.

Smoking during pregnancy can be quite harmful. It cuts the blood supply to the uterus which means a decreased supply of oxygen to the unborn child. This in turn slows the growth rate and may also harm mental development. Mothers who smoke twenty cigarettes a day are twice as likely to have a premature delivery as a mother who is a nonsmoker.

Chronic alcoholics often bear children who suffer from growth deficiency and mental retardation. Their babies may also suffer from abnormalities of the joints and heart disease. If the mother has a syphilis infection, the baby may become infected at birth, and blindness may result.

In the late sixties, a controversy erupted over the teratogenic effects of LSD. There were initial reports of central nervous system defects and malformations of the limbs caused by taking LSD during early pregnancy. There have even been some reports of defects caused by smoking marijuana during early pregnancy. However, the evidence has not been conclusive. There have been studies which suggest that moderate doses of LSD during pregnancy will not cause birth defects. There is no definite proof one way or the other. It is always better to be cautious when dealing with the life of an unborn baby. Even though there is no conclusive evidence about the teratogenic effects of these drugs, it is best not to take them during pregnancy.

It isn't difficult to understand how the mother's general health will affect the baby she carries in her womb. If a woman is starving to death, she's obviously not going to be able to supply her unborn baby with necessary nutrients. The baby will suffer for it, perhaps developing inadequately and even dying before full-term. But as we have seen, there are many more subtle ways in which the environment influences the development of embryo and fetus.

Development can be normal right up to the moment of birth and still the baby might be gravely endangered by conditions in the maternal environment. A woman with a narrow pelvic bone structure, for example, may have a passageway too narrow for the baby to pass through. As hard as the baby might push to get through, it can't do so if it is too big. In the past, babies and mothers often died in childbirth in cases like this. Today cesarean section—surgical incision in the abdomen—has become safe and an almost routine procedure, and babies can be delivered that way if a normal vaginal delivery is impossible.

It is very difficult to exaggerate the importance of the maternal environment for the developing baby.

CHAPTER THIRTEEN
Genetics and Race

***In 1921 the Congress of the United States enacted legislation restricting immigration based on national origin. This measure was motivated, in part, by a desire to keep so-called inferior racial and ethnic groups like Orientals, and southern and eastern Europeans out of the country.
***During World War II, six million Jews were slaughtered by the Nazis in the name of racial purity.
***In 1976 a bus load of young black children was attacked by angry, jeering white teenagers and adults who were vehemently opposed to school busing to achieve racial integration in Chicago.
***In the Union of South Africa, 90 percent of the population is nonwhite, but they have virtually no political rights.

Racial tensions and antagonisms continue to turn humanity against itself throughout the world. This problem has existed for a long time and cannot easily be resolved. Racial problems don't always take the extreme forms cited above. The manifestations of racism can range from distrust, to insult, to discrimination, to genocide. Racism is an emotional issue, but it is one on which the science of genetics can offer some valuable insights.

Western civilization has historically had great difficulty in reaching an objective understanding of the existence of races. In *The Biology of Race*, James C. King discusses several causes of this problem.

For one thing, our civilization was heavily influenced by Plato's concept of the *ideal type*. Plato believed that individual creatures here on Earth were actually imperfect versions of an ideal type for each species. Individuals could be measured and compared to the ideal. This ideal type did not exist in the real world. It was an imaginary creation of the mind. This approach contrasts sharply with the scientific viewpoint. Science uses accurate descriptions of what can actually be observed in the real world as its point of departure. By recording how frequently specific character traits can be observed in large populations, scientists are able to formulate a *modal phenotype*. The modal phenotype includes the characteristics most frequently found in individual members of a given racial group.

Very few serious scientists hold to the platonic approach today, but it did have considerable influence right up until just a few decades ago. Because of this influence, early students of race focused on the *differences* between races. They felt compelled to formulate ideal types encompassing these differences—even if there wasn't a single living person who resembled the ideal type. In so doing, they ignored the tremendous number of characteristics that diverse racial groups have in common. Rather than accurately describing reality, they wound up imposing arbitrary boundary lines between the races.

In 1795, J. F. Blumenbach (1752–1840) identified five human races—the Caucasian, the Mongolian, Ethiopian, American and Malay. These have been the basis for racial distinctions ever since. Blumenbach used criteria such as hair, skin, eye color and facial features to divide humanity up into these five races. To his credit, he also noted that the "innumerable varieties of mankind run into one another by insensible degrees." The scholars who followed him, however, chose to ignore this reference to the similarities of humans and concentrated their energies on the differences.

There are so many variations within each racial group that scholars found it necessary to invent subdivisions within each race. For example the Caucasians were divided into three types: the Mediterranean (southern Europe), Alpine (central Europe) and Nordic (northern Europe). Within these groups new subdivisions were also invented.

The uselessness of this kind of typology was illustrated by the publication in 1955 of an in-depth, two-volume study of the population of Ireland. The study reported that 28.9 percent of the Irish people were Nordic-Mediterranean in appearance. Or, in other words, almost 30 percent of the Irish population look like the combination of the ideal types which are supposed to be characteristic of the people living in southern Europe and Scandanavia. This study offers proof that there is tremendous variation in human populations and that traits found in groups in one location can also be found in groups in other areas.

Another source of confusion King points to is the myth of the ancient "pure" race, the Aryans. This race was said to be genetically, intellectually and morally superior to all other peoples in the world. Eventually, the Aryans met their downfall by permitting themselves to become "contaminated" through intermarriage with the inferior races around them. This myth served as a basis for philosophical justification for black slavery in America in the early nineteenth century and later for the ideology of Adolf Hitler.

It has been exceedingly difficult to discuss race without emotion. When you classify races you classify yourself, and people have usually felt a vested interest in portraying themselves in a favorable light.

The theory of white supremacy, for example, not surprisingly has been prevalent among members of the Caucasian race. Abraham Lincoln may have freed the slaves, but even he was not immune to such feelings. In the course of his historic debates with Judge Stephen Douglas, Lincoln said, "I have no purpose to introduce political and social equality between the white and black races. There is a physical difference between the two, which, in my judgment, will probably forever forbid their living together upon the footing of perfect equality; and inasmuch as it becomes a necessity that there must be a difference, I, as well as Judge Douglas, am in favor of the race to which I belong having the superior position."

It shouldn't come as a surprise that members of other races have often brought a different perspective to the question of racial superiority. Among some Indian tribes there was a legend that the Great Spirit created man by molding some clay and putting it in a kiln to bake. At first he kept it in for too short a time and it came out white and pale. On the second attempt it was burnt and blackened. On the final try, it came out a perfect copper color.

And in Alex Haley's *Roots*, we learn that on the completion of manhood training the young men of Juffure were told, "When you return home you will begin to serve Juffure as its eyes and ears. You will be expected to stand guard over the village—beyond the gates as lookouts for toubob (whites) and other savages . . ."—which gives an idea of what the Africans thought of the "civilized" white man.

As Darwinian theory became widely accepted in the late nineteenth century, it was inevitable that it would be applied to human beings. After all, weren't humans at the top of the evolutionary progression? A philosophy called social Darwinism became popular. This theory adapted the concepts of natural selection and the survival of the fittest to human society. The rich and successful were seen as the fittest people in society. Those on the bottom of the social scale were less fit.

Soon the concepts of evolution were used to examine the existence of races. Darwin had shown that species evolved and changed into new species. Man, it was believed had evolved from an apelike ancestor. Apes were black and lived in jungles. The civilization of Europe was culturally and technically more advanced than that of many nonwhite societies. This situation was seen as evidence that the nonwhite races were lower in the evolutionary ladder, more closely related to the apes than the Caucasians. The Caucasians were at the top of the evolutionary ladder, the most superior creatures developed by nature and God.

This incredible theory was accepted in scientific and nonscientific circles. It gave a pseudoscientific justification to racial injustice and European colonialism which was then in the process of dividing up the world. The Europeans could insist that they weren't merely plundering the countries of Asia and Africa for economic gain, but were accepting the white man's burden to uplift their more primitive relatives. This idea represented a total distortion of Darwin's theory. The theory of evolution was a description of a process of historical biological development. It was being transformed into a social theory that incorrectly linked evolution to preconceived ideas of morality, purity and perfection. But evolution was a process of adapting to a changing environment in order to survive. It had nothing to do with good and evil or superiority. The horse of 100 million years ago was no better or worse in a moral sense than the horse of today. But it had to change in certain ways or face extinction.

The equation of cultural level with biological evolution is absurd. Are we to conclude that the ancient Greeks were biologically and genetically superior to the Europeans of the

Dark Ages who lost their cultural heritage for a thousand years? The history of civilizations and cultures is a social history, not a biological history.

WHAT ARE RACES?

Races are the human equivalent of subspecies. There are obvious differences between races, but all races share a basic commonality. They are all *Homo sapiens.* Variation is widespread within any human population. It is often difficult to establish a firm boundary line between one racial group and another. Human variations range over a broad continuum. Though differences between certain groups might be glaring, looking at the global picture, variations tend to blend into one another.

Subspecies in nature may eventually evolve into new species separate and distinct from each other. Darwin's finches evolved in isolation from each other to a point where they could no longer interbreed. When leopard frogs from Texas are mated with leopard frogs from Vermont, most of the offspring die in the embryonic stage. Few make it to the tadpole stage. Through long periods of isolated evolution adapting to survival in divergent locales, these two frog subspecies have developed genetic programs different enough that gametes contributed by one parent from each group are incapable of producing an offspring equipped with a complement of genetic material adequate for survival.

Human races developed for millions of years in geographic isolation from each other. But there is not one shred of evidence to indicate a similar genetic differentiation between races. Despite tremendous cultural and social restrictions which have often existed, and continue to exist in many places, against interbreeding between people of different races, there is no known instance where such interbreeding has not eventually occurred. The offspring have always been normal, healthy and viable. Whatever the differences, we are more alike than Darwin's finches or Vermont's and Texas' frogs. We are one race, the human race.

Why did different races develop in the first place? Presumably in response to stresses posed by environmental conditions. The existence of widespread variation in the human race was the key to humanity's survival and cultural development. Species which become overspecialized have the most difficulty in meeting new conditions as they arise, and are the most likely to face extinction. The fact that human beings come in all sizes, shapes and colors without significant genetic differentiation is a reflection of our species' great resiliency in meeting the challenge of the environment.

It is not simple to figure out the exact relationship between the environment and the racial characteristics that evolved. Our species differs from other living things in having a unique ability to modify the environment. Humans alone are capable of shaping and controlling the environment to meet their needs. Instead of developing thick fur, as the husky did, to contend with the cold, people clothed themselves and learned to harness fire. Instead of developing tremendous strength or severely limiting the types of activity they could undertake, people discovered the principles of leverage and mechanics. They invented machines and tools to help them do work. Instead of growing long necks like giraffes to have access to sparse food supplies, people learned to farm and domesticate animals. Instead of growing flippers to swim and wings to fly, people invented boats and airplanes. If there was a selective advantage for the racial differences that evolved millions of years ago, and there probably was, the functional benefits have long been lost.

At first glance, one of the simplest traits to explain would seem to be the darker skin coloring of Africans and other groups living in regions in the vicinity of the equator. These areas are exposed to more and stronger radiation from the sun than regions farther north or south. Exposure to intensive sunlight can cause damage to skin cells. At the same time, sunlight is also beneficial in stimulating the body's production of vitamin D from substances within the body. Vitamin D is necessary for the proper utilization of calcium. Darker skin pigmentation would appear to be a selective advantage for surviving in equatorial regions. It would provide protection for the skin while permitting enough sunlight to penetrate to allow production of vitamin D.

The fact that darker-skinned populations are indeed found in areas near the equator and those with lighter skin in areas more removed from the equator lends support to this theory. But there is, as Dr. King points out, no simple correlation. The darkest Africans are found in the tropical rain forests where they are quite well protected from direct exposure to the sun. The Bushmen and Hottentots of southwest Africa live in desert areas where exposure to the sun is great. They are believed to have previously lived farther north near the equator where exposure would also have been great. However, they are much lighter in skin color than the rain-forest dwellers. In Asia, dark-skinned groups live near the equator, but they are not nearly as dark as Africans even though environmental conditions are similar. In Europe and Asia skin color lightens as you move northward. But in Asia depigmentation never went as far as it did in Europe. The inhabitants of Siberia are darker than the natives of Scandinavia and no one can offer a sensible environmental explanation for the discrepancy.

The very fact that humans can modify their environment complicates the mystery of the origin of racial differences. How can we calculate the interaction between culture and biology? A nomadic Berber in the North African desert has traditionally worn flowing capes and hoods. With this protective clothing, the Berber would certainly have experienced a different impact from exposure to the sun's rays than a naked Bushmen. But here is the dilemma. Did the customary garb of the Berbers cause them to evolve light skin, or were they impelled to adopt habits of dress that would compensate for the lack of protective skin pigmentation? Did the Bushmen develop dark skin so they could live naked, or did they live naked because their dark skin gave them enough protection to do so?

The subject becomes even more confusing when we try to analyze the relationship between other traits and the environment. The fact is that the sources of human racial differences will probably never be understood.

Race is not a burning biological question today. It is primarily a social and cultural issue. In the United States a person with any known black ancestry is still considered black. A person will be labeled black although his or her skin color may be several shades lighter than that of a person of Greek or Sicilian ancestry. Because of this social definition, he or she will be subjected to forms of discrimination that affect psychological well-being, educational opportunities and achievements, place of residence, lifestyle, job opportunities and even life expectancy. In South Africa there are stringent laws which define race and restrict the lives of blacks, whites, and Asians. In other countries, like Brazil and Puerto Rico, racial differences have far less significance in terms of how they affect a person's life chances.

The racial problems we face today stem from customs, habits, and laws put into effect by people acting in society—not from biology. There is no biological basis for racism or even for rigid racial differentiation. Despite exhaustive studies to identify a possible gene form which is found exclusively in one race or another, none has ever been found. Every possible form of gene is found in the genetic program of each race. The range of variation within a race is the same as the range between races. The only difference is the frequency with which they occur in a particular race, which translates into different modal types.

The basic underlying fact that genetics can point to is the biological unity of human beings. For hundreds of years humanity has emphasized physical distinctions that are biologically insignificant. We human beings are one race with many skin colors and many languages. We are one species.

CHAPTER FOURTEEN

Genetics and Intelligence

The idea that genes make a person a good farmer, or a good curve-ball hitter, or an automobile driver would probably seem silly to most people today. The relationship between heredity and the environment is too intricate for such simplistic explanations of human behavior to be taken seriously. But there is one human trait that many continue to believe is genetically determined, and that is intelligence. This particularly stubborn myth persists among laypeople and members of the scientific community alike. As an example, several books have been written within the last decade or so by otherwise reputable natural and social scientists arguing that blacks are genetically inferior to whites in intelligence.

Part of the problem is that the word "intelligence" means different things to different people. In common usage "intelligence" frequently has ethnocentric overtones—the attitude that one's own social grouping is necessarily better than others, that what's different isn't as good. As Dr. James King points out in his book *The Biology of Race*, in this view "intelligent" people have more or less the same lifestyle, political views, ethical values and cultural practices as the person who is making the judgment. "Unintelligent" people, on the other hand, are people who cannot see what's as clear as the nose on their face. They insist on holding ridiculous political opinions that any intelligent person would never hold. Unintelligent people don't accept or live by the same values. They are out of their minds. They live like animals.

This attitude is a very ancient human phenomenon. The soldiers of the Roman Empire were seen by Rome to be the bearers of civilization and culture. When they came in contact with other peoples, they regarded them as barbarians. Roman ways were better—of that the Romans were certain—and they had the military might to impose their rule and their ways on the people they conquered.

When Europeans roamed the world by sea in the commercial revolution that came with the Renaissance, they regarded the black Africans as subhuman. Europeans thought the blacks were too unintelligent to even have a language—they just grunted and groaned. The Europeans believed they were bringing "civilization" to the blacks—even if it meant destroying families and cultures, staging kidnappings and making slaves out of free people.

White people's relationship to the American Indian was quite similar. The "intelligent" European Pilgrims couldn't even figure out how to cope with the inhospitable New England environment in which they found themselves that first winter in the New World. They would have starved and frozen to death without the aid of the Indians. But this fact didn't stop the whites from considering the Indians ignorant savages who were inferior to the whites. This provided the pretext for the theft of ancestral Indian lands and the genocide that followed in the next two hundred years.

If we're going to get anywhere talking about intelligence and genetics, emotion and prejudices have to be put aside. We need a more neutral, scientific definition of intelligence to work with. While there isn't universal agreement on such a definition, there are few educational psychologists who would dispute the statement that intelligence is the ability to learn—to modify one's behavior on the basis of past experience. Intelligence is not purely a human characteristic. Animals, too, learn from their experience, though in different ways. Animals, too, can solve problems, though obviously not as complex as those that humans can solve. What separates human and animal intelligence is the degree to which humans can think abstractly. An animal can learn to hunt, to build a nest, or to perform tricks on command. Chimpanzees in experimental programs may be taught sign language through painstaking training, but human beings can symbolize the things they see in the world around them, remember them, record them, classify them and rearrange them and take their ability to think abstractly much, much further.

The nature-versus-nurture debate about the relative weight of genetics and the environment has been especially heated on the question of intelligence. Until quite recent decades, opinion on this controversy had historically been divided into two hostile camps. One camp held that human intelligence was a product of the environment. The human mind was a blank slate at birth, according to this view. You could teach any child anything you wanted to if you worked at it hard enough. There were no inherent differences between individuals. The opposing camp placed the emphasis on heredity. Intelligence was innate; you were born with a certain level of intelligence because of your genes. Environment had little to do with it. No matter how hard you tried to motivate a dull child to learn, nothing would be absorbed. This was the same viewpoint that attributed the successes of the Tuttle-Edwards family and the failures of the Jukes to purely genetic causes.

The generally accepted explanation of intelligence today is

a fusion of these seemingly irreconcilable theories, put forth in the 1950s by Jean Piaget. Piaget spent his later years writing on psychological topics, but his early career was spent as a biologist. His theories of intelligence reflect a strong influence of his appreciation of the developmental character of biological processes. Piaget said that intelligence is like every other human characteristic; an individual is born with a genotype which lays the basis for potential development. Intelligence is the result of the interaction between the genotype and the environment that a person grows up in. The particular kind of material that intelligence works itself out in varies from culture to culture, from place to place, from class to class.

The intelligence of an electronics engineer trained at an American university develops differently from that of an African Bushman in the Kalahari desert or an Eskimo in the Arctic. The African Bushman would do poorly if he were suddenly transplanted to live in Boston, Massachusetts, and an American engineer left on his own would have difficulty surviving in the Kalahari. Both the engineer and the Bushman would feel quite helpless in the Arctic tundra. There is a genetic component to intelligence which may set limits, just as there are genetic limits to how tall you can grow. But there is every reason to believe that an Eskimo, Bushman and American could exchange places at birth and be quite able to cope intelligently with the environment they find themselves in.

Intelligence doesn't appear full-blown at birth. Like other human characteristics, intelligence develops sequentially under genetic control. Even Einstein had to learn to master eye-hand coordination, to recognize objects, to sit up, to walk, to speak, etc., before learning more complicated skills. A four-year-old child may not understand that three pennies are less valuable than a dime. But this is not because the child is stupid or lacks intelligence. The child simply isn't old enough to understand the concepts involved in recognizing that regardless of size and quantity, the dime represents the value of ten pennies.

Even before birth, the environment interacts with the genotype to influence mental development. Chromosomal abnormalities resulting from the effects of maternal age may cause mental retardation. As pointed out in our discussion of the maternal environment, excessive smoking, alcoholism and exposure to radiation can negatively effect the development of intelligence. Events in the very first months of life are also critically important. Humans and other mammals are born before neurological development, including the brain, is completed. There is ample evidence that proper nutrition for the newborn infant is necessary for this development to be completed in a satisfactory way. It's impossible for obvious ethical reasons to conduct controlled laboratory experiences that explore the effects of malnutrition on the human brain. But experiments have been conducted on test animals. Young animals raised on a protein-deficient diet develop small brains and subsequently less competence in solving problems than animals maintained on an adequately balanced diet.

Another example of the impact of the environment in the early months of life can be seen in the fact that sensory experience is required for a healthy maturation of neurological tissue. When chimpanzees are raised in total darkness, they don't develop normal vision. Kittens and mice kept in complete darkness suffer changes in the brain cells normally associated with visual perception.

Among human beings, the relationship between parent and child is an important environmental influence on the fulfillment of intellectual potential. Psychologists have found that a strong bond of attachment between mother and child in early infancy gives the child a sense of security and encourages the development of motor skills and mental activity. These findings contradict the popular myth that a child strongly attached to its mother will be afraid to leave her side. Studies have demonstrated that a child with a strong attachment bond will feel freer to explore the environment as long as the mother is present. The infant may occasionally glance up to make sure the mother is still present but will continue to explore. When the mother leaves the room, exploration decreases sharply.

Psychologists believe that this bond is closely related to the baby's first realization that an object continues to exist even when removed from sight. This point is considered a milestone in mental development. It means that the baby is able to internally symbolize the object, which is the first step in the ability to think abstractly.

Recent studies have shown that the relationship between father and child also has an important bearing on intellectual performance from preschool to teenage years. This finding also contradicts long-standing, commonly held notions about the "negligible" role of the father in childrearing. Thus far, most of the information collected concerns the relationship between fathers and sons. In working-, middle- and lower-class families, fathers who are perceived by their sons as being warm and interested in the sons' activities and interests, tend to have sons who are brighter, more creative and imaginative. Fathers who don't seem to care about their sons' work and interests, who are perceived as domineering, often have sons who are poor problem solvers. Such children tend to see things only in broad categories and fail to grasp subtleties and lack other important analytical skills. Fathers who are seen by their sons to be restrictive, often have sons who are poor academic performers.

MEASURING INTELLIGENCE

Intelligence is believed to be measurable. Since the early twentieth century, a large number of standardized tests have been designed to gauge intellectual ability in terms of IQ. The IQ is the intelligence quotient—the ratio of the individual's score on the standardized test to the average score for his/her age group multiplied by a hundred. A person with the average score for an age group would therefore have an IQ of 100. Those above average would range higher than that figure and those below average in the other direction. Youngsters in American and British schools especially have been bombarded with IQ tests analyzing mathematical, verbal and other skills.

These tests have been quite controversial. Some psychologists feel that intelligence cannot be separated into components (like verbal and mathematical skills) but has to be evaluated as a totality. IQ tests have also been severely criticized for being heavily slanted toward middle-class experiences and values. Children from such backgrounds usually do better on these exams than children from working class and poor families. The cultural values which middle-class youngers are exposed to usually place tremendous emphasis on academic success, including the skills most directly measured by the IQ exams. These young people are influenced to do well in school, to go on to college and to enter professions. The achievement their families expect of them often goes a long way toward influencing what they actually accomplish. Because of cultural values, the availability of time and money, middle-class children are more likely to have access to trips, to museums, zoos, the country, to exposure to books, etc., than children from poorer backgrounds. These experiences have an enriching effect which will help these children do better on IQ tests.

Children from families subjected to economic and racial hardship and discrimination are frequently less likely to be motivated to be successful in academic skills. For poverty-stricken families, college and professional ambitions are often seen as being beyond the realm of possibility. A welfare family, or a family where the father earns his living as a janitor, might be quite proud of a son who grows up to become a bus driver. Whereas, a middle-class family would probably be quite disappointed in such a son. These contrasting ambitions and expectations will affect a child's performance on tests gauging academic skills. Thus the differences in IQ scores for social groupings may not reflect intelligence at all.

IQ tests have been fairly accurate in predicting future academic achievement. However, the significance of this accuracy has also been debated. If teachers operate on a preconceived notion that children are dull because of low IQ scores, what may happen is something known as a "self-fulfilling prophecy." The expectation of a low performance may make the teacher consciously, or subconsciously, fail to encourage, stimulate or challenge the youngsters to achieve the maximum possible. The end result will be the expected low performance.

In any case, it is important not to draw a simple equation between academic achievement and intelligence. Intelligence works in different ways in different situations. American Indians did not develop a written language or algebra, but they successfully developed a culture that permitted them to survive in the North American wilderness. The same is true to a more limited extent for different subgroupings within our own society. Intelligence may be directed toward development along different paths. A working-class boy may be encouraged to pursue an interest in mechanical things.

Someday he may become an excellent automobile or airline mechanic or computer technician. He may not be motivated to pursue a serious study of algebra and calculus, but this should not be seen as an indication that he doesn't have the necessary intelligence to master these subjects. Many well-educated professionals feel helpless in handling certain mechanical tasks. They don't know how to tune an automobile engine or install a lighting fixture, but this doesn't mean they lack intelligence either.

In 1977 a detailed study of IQ scores among school children in Los Angeles was released. The results graphically illustrate the influence of environment on IQ and undercut various racist theories about the genetic inferiority of blacks. The highest scoring group in the survey were students at an upper-middle-class black school. The lowest scoring group was also in a black school—this time in a poverty area. The survey showed that many of the parents of the high-scoring children had themselves been born into poverty conditions. Many had obtained a college education by taking advantage of educational opportunities offered by the GI Bill of Rights and other programs. Their children were born into a family environment that encouraged the maximum development of the intellectual potential they inherited from their parents.

CHAPTER FIFTEEN

Genetic Counseling

Rachel Leibowitz was thirty-two years old. She was pregnant with her third child and she was very frightened. Her brother's wife had just had a baby with Down's syndrome, or mongolism. Rachel had seen mongoloid children and understood that they suffered from severe mental retardation. Her brother told her that though there were many programs to help the mongoloid child take advantage of its full potential, some are so handicapped that their families eventually place them in institutions for care. Doctors had also warned her that mongoloid babies can suffer from defects of the heart, eyes and ears as well.

It broke her heart, but Rachel decided she wanted an abortion. She was afraid the same tragedy might strike her baby. "Birth defects run in families, don't they?" she said. "Maybe two normal kids are enough. Thank God we've been so lucky."

Her obstetrician insisted she consult a genetic counselor before making such a drastic decision. Genetic counselors are medical specialists who use the growing knowledge of human heredity and modern technology to fight birth defects. Rachel and her husband went to see Dr. Edward Schutta at the Brookdale Medical Center in Brooklyn, New York. Dr. Schutta understood the couple's anxieties. He deals with fear and doubts every day. Genetic counseling is as much supportive as it is informative.

The doctor explained that the first step was to determine which type of mongolism her brother's baby had—the extra chromosome or the translocated chromosome type. The extra chromosome type is the more common one. Because of a chance error the baby winds up with an extra number 21 chromosome, for a total of forty-seven chromosomes. (Figure 34) The frequency of such errors increases significantly after the mother is thirty-five years of age. For a woman in her early twenties, the odds are 3,000 to 1 against bearing a mongoloid child. At the age of thirty-five, the odds drop to 250 to 1. By age forty, the odds are down to 100 to 1. In the translocation type of mongolism, the baby has the normal number of forty-six chromosomes but extra number 21 material is attached to one of the other chromosomes. This type is not usually a random occurrence, but is an inherited family trait.

In Rachel's case, lab tests showed that her brother's baby had the extra chromosome type. Based on this information Dr. Schutta informed Rachel that her chances of having a mongoloid child were extremely slim, especially since she was under thirty-five. The deformity of her brother's baby was purely chance and had no bearing whatsoever on her own pregnancy.

But this expert advice was not enough for Rachel. She wanted to believe the doctor, but she had a nagging fear. "What if he's wrong?" To ease her doubts, the doctor ordered a chromosome study of Rachel's own blood cells. The results proved she was not a carrier of the translocated chromosome

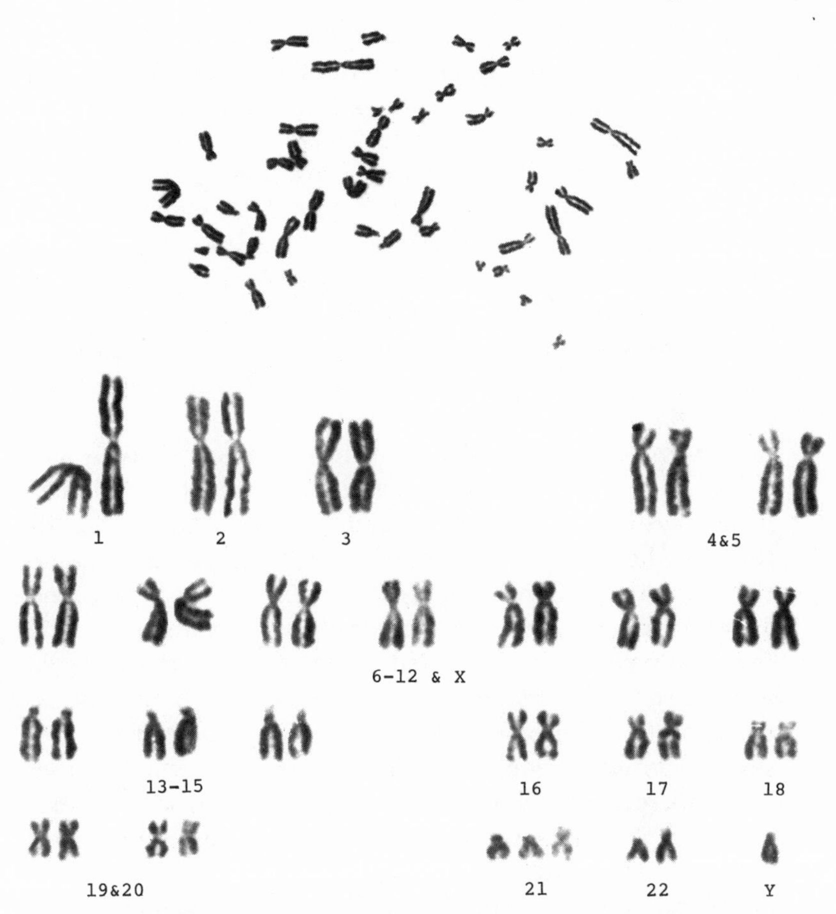

Fig. 34 Chromosomes of a male with Down's syndrome (Courtesy of the Cytogenetics Laboratory, NYU Medical Center)

type of mongolism. Her anxieties subsided. She decided to have the child.

Rachel was lucky. Twenty-five years ago, there were only ten qualified genetic counselors in the United States, and all they could do was calculate the odds and hope for the best. Today there are more than four hundred fully staffed genetic labs nationwide. They provide a full range of genetic services undreamt of twenty-five years ago. Today's genetic counselors have at their disposal accurate diagnostic tests for more than a hundred genetic disorders. They can detect more than fifty genetic conditions prenatally. Since the legalization of abortion, this means that if they wish, parents can prevent the birth of a child with severe defects. In other cases, prenatal diagnosis permits medical therapy to begin either before or immediately at birth, often before symptoms of the disorder are visible.

The greatest prenatal diagnostic tool is a relatively new and simple procedure called *amniocentesis*. After applying a local anesthetic, the obstretrician inserts a hollow needle through the mother's abdominal wall and draws a sample of the amniotic fluid in the uterus surrounding the fetus. Since this fluid contains fetal cells, a number of prenatal laboratory tests are possible. Geneticists can determine if the fetus has mongolism by checking the number and structure of the chromosomes. Chromosomal analysis also enables the baby's sex to be determined. This information is of crucial importance for parents who risk bearing children with sex-linked disorders like hemophilia. Amniocentesis wasn't performed in Rachel Leibowitz' case because at age thirty-two she was not considered sufficiently "at risk" for mongolism to warrant the procedure.

Because there are some dangers involved in amniocentesis, it is necessary to balance these risks against the statistical risks for bearing a child with a detectable abnormality. An exhaustive study on the safety of amniocentesis by the National Institute of Child Health and Human Development has drawn up some suggested guidelines. Dr. Duane F. Alexander who was in charge of the study now recommends amniocentesis in the following cases:

——For all pregnant women over thirty-five;
——For those who have already had a baby with a chromosomal abnormality;
——For those who have a family history of such disorders;
——For those known to be carriers of a sex-linked disease.

Women meeting any of these criteria should consult a genetic counselor by the end of the third month of pregnancy. This gives doctors ample time to determine whether the unborn child is normal or abnormal. It also allows the prospective parents enough time to make a decision. A recent survey of physicians revealed that 87 percent felt that amniocentesis should become a routine procedure for pregnant women over thirty-five.

Genetic counseling is becoming more and more routine. However there are still doctors who fail to refer patients to counseling when they need it. There are doctors who tell a woman who has had one child with a birth defect, "Don't worry. The odds are that the next baby will be normal." Such doctors are gambling. In one case, a doctor gave this type of advice to a woman for three successive pregnancies and each time the baby was born with the same abnormality. It turned out that she had a chromosomal abnormality in her cells which she was passing on to her children. Amniocentesis would have detected the problem early in pregnancy.

It is a sad fact that most medical schools still skimp on their genetics courses. Up until quite recently, many women found out about the existence of genetic counseling through magazine articles and not from their physicians. This unfortunate information lag is potentially disastrous for thousands of people who desperately need counseling, but don't know it exists. One young woman had a younger brother who had Duchenne's muscular dystrophy. She had watched him lose control of his muscles, wither and eventually die. She knew that the disease is sex-linked and genetically transmitted. She told her fiance she would never risk bringing a baby into the world to suffer like her brother. She refused to have children. This refusal conflicted with her fiance's strong desire to raise a family and jeopardized their marriage. But all this anguish was unnecessary.

The couple needed genetic counseling, and should have been referred to a counselor by the physician who treated her brother. Even though a man suffers from the disease, there is a 50 percent chance that his sister would not even be a carrier of the gene that causes it. Research scientists recently discovered a new method to test a woman for carrier status, but this is not yet in use. Even if she were a carrier, any baby girl she delivered would not suffer from the disease. Since the baby's sex can be determined prenatally through amniocentesis, she would know in advance if the baby would be male or female.

THE DOCTOR-PATIENT CONTROVERSY

The relationship between the genetic counselor and the patient is a very special one. The issues involved are too complex, too emotional, too intimate for the doctor to stand as the authoritative decision-maker. The old "You're-the-doctor-whatever-you-say-goes" attitude is not applicable. The geneticist can only present the facts; the parents must decide what to do.

"When you have a broken arm, everybody knows you've got to get it fixed," says Dr. Schutta. "But with genetic counseling the issues are not so clear. Whether to have children, or not to have children, whether a particular risk for abnormality in a child is too great, or not too great—these are decisions that do not have a right and a wrong to them."

There are some geneticists who disagree with Dr. Schutta. They believe that there are simple right and wrong decisions.

They believe that the genetic counselor should do whatever is possible to make sure that children with birth defects are not born. This disagreement on the doctor-patient relationship has erupted into a heated debate in the last few years.

A paper presented at an important national genetics conference in San Francisco in June 1978 complained that too many parents were deciding to have babies even though they knew from genetic counseling that their babies would be born with birth defects. The author called on genetic counselors to take a more "directive" approach.

Other geneticists argued that there is no such thing as a single, unequivocal, "rational" choice in an area as subjective as the decision to have a baby or an abortion. What is a rational choice must be defined within the context of the family involved. What is rational for one family is out of the question for another. An informed decision-making process is in fact "rational" regardless of the decision reached by the parents. The decision to have a baby or not must be made by the parents, based on the medical information available and the parents' own understanding of themselves and their families. A decision can be rational even if it isn't the same one that a geneticist would reach under similar conditions. What is important is that the parents be able to make the decision that they can live with and that the genetic counselor support them in whatever path they choose.

How these two contrasting views of the role to be played by the genetic counselor are put into practice raises important ethical questions. Not all abnormalities carry the same negative implications. Particularly in regard to sex chromosome abnormalities, the *way* that a counselor describes the situation may determine or influence the decision the parents make. For example, babies who have the **XO** genotype mentioned in Chapter Six characteristically have a condition known as Turner's syndrome. They are inadequately developed females. They tend to be short when they grow up—

perhaps four-feet-ten-inches as a maximum. They *could* have webbed necks and occasionally suffer from *repairable* defects of the heart and kidneys. They cannot bear children and have inadequate secondary sex characteristics development. Hormone treatments, however, can assure more normal breast and hip development. Turner's babies do not suffer from mental retardation or physical handicap.

Should the counselor merely explain the facts and let the family mull it over? If they decide to have the baby, should the counselor be supportive and assist in making arrangements for necessary therapy for the child? Or, should the counselor emphasize the opinion that it is terrible to bring an abnormal child into the world, knowing that when the child grows up she will never be able to have children of her own? Should the counselor stress the expense of corrective surgery? In other words, should the counselor pressure the parents not to have the baby?

Which is the ethical approach? This is a question that concerns society in general, and not just geneticists. Genetics now has at its disposal knowledge and technology which can help the human race. This knowledge is not the private property of specialists, but an acquisition of society as a whole. How it is used is a social decision.

There are some like Washington, D.C., physician Richard M. Restak who fear that genetic counseling holds dire consequences for the future. They fear that civilization will become impatient with people who suffer from genetic disease that could have been detected before birth. Society might feel that such people should never have been born. Compassion and sympathy for the afflicted might disappear.

This fear does not seem well-founded. No amount of tampering with the gene pool could possibly eliminate birth defects forever. Even if hemophilia carriers never have children, hemophilia will never disappear. Mutations occur randomly all the time. And there is no way to tell whether any mutation, including the hemophilia mutation, has occurred until it shows up in the phenotype—until a hemophiliac is born. So science isn't going to stop hemophilia. But what genetic counseling can do is help families in which the mutation crops up keep other relatives from suffering the disease.

The same holds true for other genetic disorders. Even if amniocentesis is performed on all women over thirty-five, mongoloid children will still be born. Amniocentesis is not recommended for younger women for safety reasons and the high odds against mongolism. But once in three thousand times, a twenty-year-old woman will give birth to a baby suffering from Down's syndrome. Genetic counseling and amniocentesis can greatly reduce the frequency of mongolism, not make it disappear. People are not impatient with the victims of genetic disease because of genetic counseling. The fact is that society has historically been impatient with the maimed, the crippled, and the handicapped. It has tried to forget about these people by putting them in institutions and freezing them out of full participation in social life. It has blocked them from reaching their full potential as human beings. We have made progress in the detection and prevention of these disorders. But social attitudes, support systems, and medical and educational facilities have deplorably lagged behind the technological advances seen in genetics. This lag shouldn't trigger an attack on genetic counseling; it just emphasizes the need for social attitudes to improve.

Another serious question raised about the counselor-patient relationship is whether geneticists give their patients more information than they can handle. Parents are often too upset to absorb the information being explained to them. A study at a Philadelphia hospital revealed that only 19 percent of the patients participating in a genetic counseling project could explain what they had been told. Having a little bit of improperly understood information can sometimes be worse than no knowledge at all. Sometimes parents who learn they are carriers of a defective gene are overwhelmed by feelings of guilt, even though they have nothing to feel guilty about. The genetic program we inherit is something totally beyond our control. There is no way to avoid a defective gene passed on from previous generations or to avoid a randomly occurring mutation. A person doesn't carry a defective gene because he or she has done something wrong. Genetic disease should carry no social stigma, but it often does.

There is evidence that people do indeed have a difficult time handling the information given them by genetic counselors. Sometimes spouses become angry at each other. Divorce rates for families with genetic disorders seem to be higher than average. However, this seems not so much an argument against genetic counseling as much as it is an argument for better, more thorough, more supportive counseling. The counselor not only has to inform the patients, but must also help them adjust to and live with the information.

The relationship between genetic counselors and their patients is a very complicated one and one that is still in the process of being defined. It involves a subtle balance between conveying necessary information and helping the patient live with the consequences of that information. The clarification of that relationship is the topic of discussion today and will continue to be so in the years to come.

CHAPTER SIXTEEN

Genetics and the Future

Genetics has come a long way since the early part of this century. The genetic code has been broken. We understand a great deal about how traits are inherited. We understand how DNA duplicates itself, how it is structured and how it changes. We understand the nature of the interaction between our genes and the world around us. We can detect some birth defects even before birth. We can even tell the sex of an unborn child still snugly sheltered in its mother's womb.

What does the future hold for genetics?

Despite the progress, many questions remain unanswered. No one has yet seen a gene, and we don't understand the mechanisms by which cells signal each other in carrying out the genetic program. Research into these questions is going on right now and will continue in the future.

Genetic counseling programs and their supporting laboratory services will increase in importance. Today there is more awareness of genetic counseling than there was five or ten years ago, but it is not yet a household word. In the future, genetic counseling will become a familiar and routine procedure for thousands of American families.

Research in human genetics is intimately associated with the clinical practice of genetic counselors and their labs. Very few human genetic disorders can be studied in laboratory animals. The overwhelming majority are found only in human beings. It is quite likely that the majority of human chromosomal disorders have not even been identified yet. New data are really accumulated on a case by case basis as individual physicians encounter patients with previously unrecorded chromosomal abnormalities. New and improved laboratory techniques permit closer examination of the chromosomes and detection of problems previously unnoticed. In the last few years there has been a trend toward the establishment of centralized birth defects registry agencies, operated by government or private health care facilities. These agencies record, sift and synthesize the information coming in from isolated and dispersed sources, as well as make the information readily available to genetic counselors. This trend will intensify in the years ahead.

Because its subject is the very basis of life, genetics will be a source of heated controversy in the coming decades. When geneticists limited themselves to pea plants and fruit flies, their work sparked far less concern. Now that we have learned enough to apply our knowledge to human beings, people are more easily aroused. Clearly, important issues of ethical, moral and social significance must be thrashed out openly and honestly for genetics to serve humanity responsibly in the future. Unfortunately, the issues sometimes get clouded in emotionalism and distortion in the mass media. This makes the necessary societal discussion that much more difficult to accomplish.

GENETIC ENGINEERING

One issue that has sparked considerable controversy and distortion is genetic engineering. Genetic engineering is the manipulation of genetic material for specific purposes. It has actually been practiced by humankind even before reproduction and heredity were understood. Manipulation of plant reproduction to yield better and sturdier crops is a form of genetic engineering that goes back 6,000 years. Artificial selection of domestic animals is another example. No one ever found a moral dilemma posed by the farmers' quest for better beef cattle. In fact, these types of genetic engineering have enhanced our ability to survive by adjusting the food supply to human needs. But when genetic engineering started being applied to human beings, important ethical questions were raised—and quite correctly. The most important of these issues were mentioned in regard to genetic counseling in Chapter Fifteen.

In 1978 public attention turned to another dramatic example of genetic engineering—cloning. Cloning is the use of a single cell from an adult to produce a genetically identical individual. Early in the year a controversial book was published which purported to tell the story of a human baby cloned from a single skin cell of an adult male. This report was so hotly disputed that the publisher declined to list the book as a work of "nonfiction."

The mass media picked up on the cloning controversy. In March, one of the sensational national tabloid newspapers ran an article headlined "Babies Made to Order" which alleged that geneticists could produce babies with the looks of a Hollywood actress and the "brains" of Henry Kissinger by rearranging pieces of genetic material taken from such celebrities and using this material as the basis of a clone. The article even raised the specter of "bringing back well-embalmed historical figures—for instance, trying to clone from cells scraped from the mummy of Egyptian King Tutankhamen."

Unsubstantiated accounts and hysterical stories like these may sell newspapers but they contribute nothing to the public's understanding of genetics. Intelligence is not the result of a single piece of genetic information; nor are "good looks." Ideas like this completely obscure the delicate interrelationship between the genes and the environment that makes people what they are. And as for the idea of bringing King Tut back, it is just incredible! The idea that dead cells or nonliving matter could produce life was discredited long, long ago. Living matter alone has the ability to reproduce itself.

Cell-culturing techniques have successfully used the principles of cloning for many years—but not to create new organisms. A sample blood or skin cell has been taken from a patient and placed in a culture medium, maintained and fed until it begins to divide and produce cells identical to itself. These cells can then be studied and analyzed. This type of cloning has never provoked an outcry.

What the media have focused on is the forming of a new organism identical to the parent organism from a single cell taken from the parent. The sequential functioning of the genetic program makes this type of cloning difficult to achieve. As differentiation progresses, certain parts of the program become deactivated in particular cells because there is no longer any need to draw on that information. Thus a skin cell might not carry an activated form of the genetic material responsible for the formation of internal body organs. In adult organisms, differentiation has obviously been completed. You might be able to clone other skin cells from a single skin cell, but trying to clone an entire organism would pose serious problems. Scientists have had some success in cloning experiments with amphibians such as frogs and toads. A clone was successfully made from a single cell taken from new embryo at an early stage of development before any genetic material would have been deactivated. The frog created from this clone, however, was not identical to an adult parent, but a twin to the embryo—which is different from the cloning which has provoked so much controversy.

There has been some success in transplanting the nucleus of a cell taken from an adult frog's intestine into a fertilized egg cell which has had its nucleus removed. This would produce an organism genetically identical to the parent. But even in successful cloning cases, it has to be stressed that no two individuals are ever exactly identical. The genotype-environment interaction applies to identical twins and clones as well. Something as seemingly trivial as how and where identical twins lie in the amniotic sac during pregnancy can cause differences in physical appearance. The same possibility for variation in the phenotype exists for clones also.

Critics of cloning occasionally invoke the *Brave New World* image and warn against the creation of docile workers and brutal soldiers in a future totalitarian world where geneticists would control who mates and reproduces. The view implicit in these warnings is reminiscent of the mistaken theories which equated behavior patterns and intelligence with the inheritance of genes. Of course the evidence is that such traits and behaviors are socially conditioned. Adolf Hitler used techniques of propaganda, psychological manipulation and violent intimidation to control the behavior of masses of people. The people who obediently committed suicide at the People's Temple in the Guyanese jungle in November 1978 were conditioned to behave in an unthinking manner. Nazi party members and religious cultists cannot pass their acquired characteristics on to their descendants through their genes. There is no biological reason to assume that their clones would necessarily behave in the same way. Cloning isn't necessary to create a totalitarian society. A totalitarian system, with or without clones, is something that people committed to freedom must combat.

Experimental work with *recombinant DNA* is another aspect of genetic engineering that received media attention recently. Scientists have attempted, with some success, to isolate, identify and recombine segments of DNA with DNA in existing organisms. Recombinant DNA offers future possibilities that include the controlled production of food and usable energy from waste materials. Geneticists also hope to use it to produce hormones, enzymes and antibiotics necessary for medical purposes. In the far distant future there is the possibility that recombinant DNA may be used to replace defective genetic material within living organisms—to actually replace defective genes which cause birth defects!

Recombinant DNA also poses certain dangers, including the creation of new life forms which might interfere with the balance of nature developed over millions of years of evolution. The General Electric Company has already utilized recombinant techniques to create a new type of bacterium which feeds on oil, and could be used to clean up oil spills. These bacteria would be released literally to eat up the oil. But releasing a new organism in the environment is fraught with danger. What effect will the presence of this new organism have on fish and plant life in the ocean? What will happen to sea life that might consume it? What will happen to the bacterium after it consumes the oil? Will it disappear or will it gradually evolve, and if so, in what direction? How will it affect the balance of nature?

In the years ahead, guidelines will have to be formulated

to safeguard our world against the hazards posed by this kind of research.

People have been trying to figure out the mysteries of heredity for over 2,000 years. Now we have finally reached the point where enough is understood so that we can actually begin to *apply* what we know. We can actually do something to spare families and babies from the heartache and suffering of serious genetic disease.

It is only natural to be deeply concerned about the possible abuse of the knowledge we now have, or will gain in the future. There is no denying the serious ethical and moral questions opened up by breakthroughs in genetics, past, present and future. The ethical and moral questions have already posed themselves in genetic counseling on an everyday basis. These issues are real enough and must be dealt with. These issues are not the exclusive property of scientists. They constitute social questions and the solutions must meet humanity's needs. To deal with these problems we need an informed society. As complicated as the subject may be, it can be understood by ordinary people. We cannot afford to be carried away by science-fiction-like hysteria manufactured by the mass media to sell newspapers. The stakes are too high to permit sensationalism to jeopardize the progress that has been made and could be made in the future. It is hoped that this book has helped the reader gain some insight into the state of genetics today.

Glossary

Amino acid Organic molecules which are the building blocks of protein. Some amino acids are produced within the cells. Others must be obtained from the environment.

Cell The basic unit of life, cells are the smallest living things that can feed, grow, and reproduce independently.

Chromosome A chain of genetic material coded in DNA in the cell nucleus. DNA assumes this squat, rodlike chromosome appearance only during cell division. At other times it appears as diffuse threads called chromatin.

DNA Deoxyribonucleic acid. The chemical substance which codes the genetic information.

Embryo An animal or plant in the early stages of development. In humans, the developing baby is referred to as an embryo until the end of the eighth week of pregnancy.

Fetus An unborn animal in the later stages of development in the uterus. In humans, the developing baby is referred to as a fetus from the beginning of the ninth week of pregnancy until birth.

Gametes The reproductive cells (sperms and egg cells) of sexually reproducing organisms containing half of a full complement of the genetic program.

Gene A segment of the chemical DNA which carries a basic unit of hereditary information in coded form.

Gene pool The full range of genes existing in a species population which may contribute to the genetic makeup of the next generation.

Genetic program The full set of genetic information coded in DNA. The program governs and controls development and functioning of living organisms.

Genotype The genetic makeup of an individual living organism. In sexually reproducing organisms, half of the genetic material is contributed by each parent.

Hybrid In plant or animal reproduction, the offspring of two organisms pure for contrasting trait(s).

Induction In embryology, the process by which one layer of cells induces or causes changes in the layer of cells next to it. It is one of the means by which an embryo develops.

Meiosis Cell division that produces reproductive cells carrying only a single set of chromosomes.

Mitosis Cell division within a living body that produces daughter cells similar to the mother cell.

Molecule The smallest unit into which any substance can be divided and still retain its original characteristics.

Mutation The source of all genetic variation. It is the sudden change in the genotype of any organism. Its impact is felt through a change in the production of protein.

Nucleotides The small organic molecules which are the basic units of DNA. Nucleotides are composed of a sugar, a phosphate group and a base. The bases are of four types: adenine, cytosine, guanine and thymine. A sequence of three nucleotides specifies one of the amino acids. The arrangement of these nucleotide triplets specifies the protein to be constructed from the amino acids.

Phenotype The physical appearance and behavior of an organism.

Polygenic The mutual interaction of a number of different gene pairs to cause specific traits in an offspring.

RNA Ribonucleic acid, the general name of three substances which are used to transcribe and translate the genetic code contained in the DNA.

The substances are messenger RNA (mRNA), ribosomal RNA (r-RNA) and transfer RNA (tRNA). They all appear as chains of nucleotides which are modeled after corresponding strands of DNA. mRNA carries the coded genetic message from the cell nucleus to the cytoplasm outside the nucleus. Translation of the message carried by mRNA takes place at cell substructures called ribosomes, the major component of which is r-RNA. tRNA brings amino acids in the sequence specified by the mRNA to the ribosomes in order to produce the proper proteins.

Species The smallest group into which most living things are divided by means of having characteristics in common. Among the principal characteristics that define species is that members can breed within the group but not outside it.

Bibliography

Asimov, Isaac, *The Genetic Code*, 1962, Signet Science Library Books, New American Library, New York.
Asimov, Isaac, *A Short History of Biology*, 1964, The Natural History Press, Garden City, New York.
Barish, Natalie, *The Gene Concept*, 1965, Van Nostrand Reinhold Co., New York.
Basler, Roy P., *Abraham Lincoln: His Speeches and Writings*, 1946, The World Publishing Co., Cleveland and New York.
Bergsma, Daniel, Editor, *Contemporary Genetic Counseling*, 1973, The National Foundation-March of Dimes, White Plains, New York.
Bergsma, Daniel, Editor, *Symposium on Intrauterine Diagnosis*, 1971, The National Foundation-March of Dimes, White Plains, New York.
De Grouchy, Jean and Catherine Turleau, *Clinical Atlas of Human Chromosomes*, 1977, John Wiley and Sons, New York.
Dennis, Wayne, Editor, *Readings in Child Psychology*, 1963, Prentice-Hall, Englewood Cliffs, New Jersey.

Fraser, George and Oliver Mayo, *Textbook of Human Genetics*, 1975, Blackwell Scientific Publications, Oxford.

Goldstein, Philip, *Genetics Is Easy*, 1961, Lantern Press, New York.

Hutt, Frederick B., *Animal Genetics*, 1964, The Ronald Press, New York.

King, James C., *The Biology of Race*, 1971, Harcourt Brace Jovanovich, New York.

McKusick, Victor A., *Human Genetics*, 1969, Prentice-Hall, Englewood Cliffs, New Jersey.

Moore, Keith L., *The Developing Human—Clinically Oriented Embryology*, 1977, W. B. Saunders Company, Philadelphia.

Raab, Carl and Joan Raab, *The Student Biologist Explores Genetics*, 1977, Richards Rosen Press, New York.

Roberts, J. A. Fraser, *An Introduction to Medical Genetics*, 1970, Oxford University Press, London.

Scheinfeld, Amram, *Your Heredity and Environment*, 1965, J. B. Lippincott, Philadelphia.

Smith, John Maynard, *The Theory of Evolution*, 1958, Penguin Books, Baltimore.

Stevens, Joseph H., Jr., and Marilyn Matthews, Editors, *Mother/Child, Father/Child Relationships*, 1978, National Association for the Education of Young Children, Washington, D.C.

Sturtevant, A. H., *A History of Genetics*, 1965, Harper and Row, New York.

Waddington, C. H., *How Animals Develop*, 1962, Harper Torchbooks, New York.

Wadsworth, Barry J., *Piaget's Theory of Cognitive Development*, 1971, David McKay, New York.

Watson, James D., *The Double Helix*, 1964, Atheneum, New York.

Weiss, Paul, *Principles of Development*, 1939, Henry Holt and Company, New York.

Index

About the Authors

Sandy Bornstein graduated from Barnard College and holds a master's degree in human genetics from McGill University. She has worked in the medical genetics program at Brookdale Medical Center and is currently chief technician at the New York University Medical Center genetics laboratory. Her husband, Jerry, graduated from New York University and is a freelance writer who has had a number of articles published in newspapers and magazines. The Bornsteins live with their two daughters in a renovated brownstone house in New York City.